Physical Fundamentals of Oscillations

Leonid Chechurin · Sergej Chechurin

Physical Fundamentals of Oscillations

Frequency Analysis of Periodic Motion Stability

 Springer

Leonid Chechurin
Lappeenranta University of Technology
Lappeenranta
Finland

Sergej Chechurin
St. Petersburg State Polytechnic University
St. Petersburg
Russia

and

St. Petersburg State Polytechnic University
St. Petersburg
Russia

ISBN 978-3-030-09160-6   ISBN 978-3-319-75154-2   (eBook)
https://doi.org/10.1007/978-3-319-75154-2

Printed on acid-free paper

This Springer imprint is published by the registered company Springer International Publishing AG
part of Springer Nature
The registered company address is: Gewerbestrasse 11, 6330 Cham, Switzerland

Acknowledgments

The authors would like to acknowledge Dr. Elena Kutueva for her tremendous help in translating the book into English and editing the main part of the book. Without her, the writing would have taken much longer if not be possible at all. We thank Mrs. Aleksandra Tcaregorodtceva for the cover design.

The cover photo depicts a scene from a family anecdote—a father-and-son story. The tale begins with the first book on period motion stability analysis published by Sergei Chechurin, in Russian, almost 40 years ago. The book had a silhouette of a boy standing on a swing on its cover. The illustration was described: "It was a summer morning on the bank of a lake near our house. I was teaching my three-year-old son how to use the old wooden swing crafted by my father. My plan was to lecture how to generate stable self-oscillations, in other words, to swing. I explained a piece of theory and gave the swing an initial deviation. The little boy tried his best, stood up and squatted down, but no parametric resonance emerged. After several attempts, I gave up and pulled him off the swing. We were clearly angry with each other, and a healthy youngster's cry exploded across the morning lake. My wife, Nina, immediately appeared. Without further ado, the child was placed back on the swing, and without any calculations, I was given an exact upper boundary for the magnitude of forced oscillations. Much later, I realized how unfair we could be to children. Of course, I was wrong. Due to his small height, h, Leonid could not excite parametric resonance at all. The approximate excitation condition is $h > 4l/N$, where l is the suspension length, N is the number of free oscillation swings with small deviation and a fixed child on the seat." Many, many years passed and little Leonid grew. He studied hard and gained two degrees in control theory and became a co-author. And a father himself. The cover picture shows the next generation, Leonid's two sons: the younger son, Peter, is doing simple forced oscillations, while his brother, Sergei, is enjoying parametric resonance. Can the thoughtful and attentive reader of the book tell if these are first or second parametric resonance oscillations?.

Annotation

The book is supposed to be useful for engineers, graduate students, and scientists that specialize in designing dynamic systems in different branches of engineering. The book can also be useful for the researchers that deal with the study of dynamic processes in relatively new fields: macroeconomics, biology, etc.

The simple and visual presenting makes the book useful as a teaching material for the courses on general physics, theoretical mechanics, theories of electrical engineering, radio engineering, automatic control, economics, etc.

Contents

Introduction

The realm of oscillations is enormous. It involves the oscillations in nature, society, technology, and economics. Oscillation processes occur within chemical and biological environments, etc. There is a wide variety of oscillations. For example, mechanical system oscillations are different from those in electric and radio circuits by their physical nature, and they differ together from oscillations in liquids and gases and all the more in socioeconomic spaces.

In spite of a wide variety, all oscillations have the same mathematical basis because they are described by differential equations such as linear and nonlinear, time-invariant or time-variant, steady and unsteady, and ordinary and partial. We could call the equations of Mathieu and Hill as the first study of time-variant systems. There are a number of remarkable contributions of Soviet scientific school: N. Krylov and N. Bogoljubov (the theory of nonlinear oscillations, 1937–1943), L. Mandelshtam and N. Papaleski (parametric oscillations in RLC circuit, 1947), M. Levinstein (parametic oscillations in electric machines, 1948), P. Kapitsa (inverted pendulum stabilization, 1948), A. Lyapunov (stability theory, 1892), B. Bulgakov and A. Andronov (oscillation theory, 1954–1959), V. Bolotin (elastic system stability, 1956), V. Taft (spectral theory for circuits with variable parameters, 1964), V. Fomin (parametric resonance in distributer parameter systems, 1964), V. Yakubovich and V. Starzhinski (differential equations with periodic coefficients, 1972), E. Popov (allied theory for nonlinear system control, 1973), E. Rozenvasser (periodic nonstationary control systems, 1973), and many others. We place here the books that established a school from which the presented study was built.

The results of S. Chechurin obtained in 1977–1983 formed the first monograph on the frequency analysis or parametric resonance [9]. The study was developed in the book [23]. The joint efforts of authors on application of this approach to the analysis of the wide class of time-variant and nonlinear systems resulted in the book [22] in 2005. Having been published long time ago in small circulation in Russian language, these two books should be counted as the roots for the present edition, largely changed, developed, and targeted at wider audience.

In addition to a common mathematic basis, oscillations also possess a common physical basis which is expressed in terms of physically measured oscillation parameters such as frequency, phase, and amplitude. In an explicit form, the parameters appear in the frequency methods of a dynamic system analysis. A frequency analysis is well known in its general form as a spectral method by V. Taft or generalized Hill's method (see *Taft V. A.*, Spectral methods of time variant circuits and system modeling (in Russian), Moskva, 1978, and also *Randall R. B.*, Frequency analysis, B&K, LTD., 1987. We also need to acknowledge most the classical, based on mathematical methods, study of nonlinear dynamics in frequency domain by G. Leonov [24]). In an elementary single-frequency approximation, the method is known as a first-harmonic method, or harmonic balance method, or harmonic linearization method, etc.

This book has several essential features in contrast to other oscillation studies. The first difference is the wide range of analyzed dynamic systems. We deal with linear and nonlinear, steady and unsteady, and discrete and distributed parameter systems. The second feature is the development of describing function method for the class of periodically unstable dynamic systems. The third feature is the use of describing function derivative to analyze the loss of stability in nonlinear systems. This enables to perform a "hidden" analysis using a single-frequency approach to obtain results that are unreachable with the first approximation. Lastly, there is the main feature to integrate the total work. It is the use of a unified approach in solving physical problems. The key point of the approach is based on a common balance principle between phases and amplitudes of dynamic system oscillations, and it is given physical and graphical interpretations. The text is illustrated by numerous simple examples of oscillating system analysis and its numerical simulations. It is significant that all the results are achieved without solving differential equations, and all the conclusions and experiments are reproducible by the reader with the basic skills in popular mathematical packages (e.g., Simulink and MATLAB).

Since amplitude, phase, and frequency are the basic physical parameters of oscillatory motion, the first part of the book deals with amplitude–phase–frequency characteristics or *hodographs* or *Nyquist plots*. Although the reader could easily find more detailed information on this in control theory textbooks, the brief summary of Part I emphasizes the physical aspects of the characteristics and also gives the key for better understanding of the remainder of the text. Part II introduces the original stationarization method with respect to periodically nonstationary dynamic systems. Parts III and IV apply the results obtained in Part II to approach the problems of oscillation stability evaluation in nonlinear systems and parametric oscillation control. Part VI provides methods to increase the accuracy of the single-frequency approximation used in the book. The rigorous solutions of robust design problems to both nonstationary and nonlinear systems are given in Chap. 16.

The text is illustrated by a wide spectrum of applied problems from technology and economy (Part V). The authors of the book are experts in the field of theory and engineering of control systems only. They obviously cannot pretend to know more precise or more rigorous models for oscillatory systems available for professionals in the specific fields. The aim of the book is to supply the professionals with research procedures, especially because the truth ultimately comes from the experiments that are available for the professionals. Thus, the illustration of the nonfrequency conditions for parametric resonance excitation is given on mathematical models only in the book. These mathematical models may or may not have a link of physical real systems. We also use known models of stationary systems that are used as an approximation for nonstationary ones, assuming small variation of parameters and coordinates. It should be stressed that the conditions for parametric resonance excitation in time-variant system derived by the models of periodically nonstationary or stationary system are the same very often (see Sect. 16.6).

The authors hope that the bridge between the oscillation theory and control theory will help to improve the oscillation control quality as well as to avoid catastrophic consequences of parametric resonance excitation.

The annex presents the correction of the one-frequency approximate analysis of periodic time-variant and nonlinear systems. The correction takes into account the influence of certain higher harmonic components. The final chapter presents the sufficient assessments of one-frequency approximations derived by robustness analysis.

The book is written for engineers, graduate students, and scientists who design and analyze dynamic systems in different branches of engineering. The book can also be useful for the researchers who study the dynamics of complex systems in macroeconomics, biology, etc.

The simple and visual exposition allows the use of book to a variable extent as a learning aid for the courses of studies on general physics, theoretical mechanics, theories of electrical engineering, radio engineering, automatic control, economics, etc. We also recommend books [1–3] for the beginners as a pre-reading.

Part I
Amplitude–Phase–Frequency Characteristics of Linear Steady-State Systems

The introductory Part I presents the modeling methods and basic properties for linear dynamic systems with time-invariant parameters. Ordinary linear differential equations serve as the mathematical basis for continuous dynamic systems. Finite difference equations are used for modeling of discrete-time systems or for digital modeling of distributed parameter systems. The language of transfer functions and frequency characteristics spoken by many engineers is used in the final representations of the dynamic systems.

Chapter 1
Continuous Systems

1.1 Transfer Functions of Dynamic Systems

Let us take the differential equation in its operator form as a base definition of the dynamic system

$$G(s)x_{out}(t) = H(s)x_{in}(t), \tag{1.1}$$

where $x_{out}(t)$ is the output coordinate, basically the coordinate/motion we are interested in; $x_{in}(t)$ is either the input coordinate or external action, disturbance, signal, etc.; $s = d/dt$ is a time differentiation operator; $G(s)$ and $H(s)$ are the operator polynomials of n th and m th orders:

$$\begin{aligned} G(s) &= g_n s^n + g_{n-1} s^{n-1} + \cdots + g_1 s + g_0 \\ H(s) &= h_m s^m + h_{m-1} s^{m-1} + \cdots + h_1 s + h_0, \end{aligned} \tag{1.2}$$

where g_i and h_i are constant polynomial coefficients called dynamic system parameters. Polynomials (1.2) are reduced to their standard form by factoring out one of the marginal (either leftmost or rightmost) parameters. If the parameters g_0 and h_0 are factorized out, the parameters g_i and h_i take the dimension of $(sec)^i$, so all the polynomial summands become dimensionless taking into account the s operator dimension of $(sec)^{-1}$. We assume $n \geq m$ which means that the system has certain inertia (delay) in its input–output signal transmission; in other words, the system can be physically realized. This type of transfer functions is called *proper*.

Equation (1.1) can be rewritten as the polynomial ratio:

$$W(s) = \frac{X_{out}(s)}{X_{in}(s)} = \frac{H(s)}{G(s)}, \tag{1.3}$$

which is called a *transfer function* of the system. In contrast to notation (1.1), X_{out} and X_{in} in (1.3) are not the coordinates themselves but their Laplace transforms. The

© Springer International Publishing AG, part of Springer Nature 2017
L. Chechurin and S. Chechurin, *Physical Fundamentals of Oscillations*,
https://doi.org/10.1007/978-3-319-75154-2_1

transfer function denominator is a characteristic polynomial of the system, and when it is equal to zero, the equation is called *characteristic equation*. The characteristic equation roots are called *poles* and the numerator polynomial roots are called *zeroes* for the transfer function.

A direct analysis and especially an experimental study of system (1.1) are often difficult or even not possible for some reasons such as cumbersome calculations, unavailability of output coordinate measurements, and initial system instability. On similar occasions, it becomes reasonable and sometimes necessary to study another system modified with known properties and performance which allow the determination of those concerning the initial system. Such an approach is most often realized by either introducing or removing a number of connections in an initial system. The feedback imposition method found the most practical application as usual to stabilize an initial dynamic system. For example, the control theory is based on that method. The principle of disconnections or removing connections known as an equivalent generator method has been used in the theory of electric circuits for a long time [2].

Let us introduce the negative feedback or output-to-input connection to system (1.1) as

$$G(s)x_{out}(t) = H(s)\left[x_{in}(t) - x_{out}(t)\right].$$

Then, we obtain the new closed-loop system with a description

$$[G(s) + H(s)]\,x_{out}(t) = H(s)x_{in}(t) \tag{1.4}$$

and a transfer function

$$W_{c\ell}(s) = \frac{X_{out}(s)}{X_{in}(s)} = \frac{W(s)}{1 + W(s)} = \frac{H(s)}{G(s) + H(s)}. \tag{1.5}$$

The new system is evidently connected with previous open-loop system (1.1) through the transfer function (1.3) but it has different descriptions, different properties, and different characteristic equations

$$D(s) = G(s) + H(s) = 0. \tag{1.6}$$

Equation (1.6) does not depend on an input disturbance as well as the input location. It defines all the properties of the autonomous closed-loop system $(x_{in} = 0)$. The knowledge of either (1.3) or (1.5) transfer function is important because it allows finding the unknown one. Thus, from (1.5)

$$W(s) = \frac{W_{c\ell}(s)}{1 - W_{c\ell}(s)}. \tag{1.7}$$

Fig. 1.1 Elementary
closed-loop system example

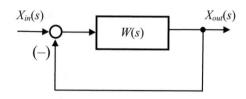

Fig. 1.2 Block diagram of
differential equation system

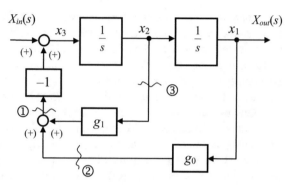

The transfer functions are used to represent the dynamic systems and their units as block diagrams. Figure 1.1 shows the block interpretation of an elementary closed-loop system.

Knowing the block representation, the system transfer function is derived as a result of simple structure transformations. Thus, the transfer function of consequentially connected units is the product of unit's transfer functions and the transfer function of parallel unit connection is its sum. Transfer function (1.5) describes the feedback connection.

Figure 1.2 shows the block diagram of a dynamic system which is described by the second-order equations

$$\dot{x}_1 = x_2$$
$$\dot{x}_2 = x_3$$
$$x_3 = x_{in} - g_0 x_1 - g_1 x_2.$$

This block diagram is also called as a diagram of state variables, x_i.

The transfer function of the system takes the form

$$W_{c\ell}(s) = \frac{X_{out}(s)}{X_{in}(s)} = \frac{X_1(s)}{X_{in}(s)} = \frac{1}{s^2 + g_1 s + g_0}.$$

The transfer functions are derived by breaking the connections 1, 2, and 3 one after another as follows:

$$W_1(s) = -\frac{g_1 s + g_0}{s^2}$$

$$W_2(s) = -\frac{g_0}{s(s + g_1)}$$

$$W_3(s) = -\frac{g_1 s}{s^2 + g_0}.$$

Apparently, the transfer functions are different, but in all cases the difference between the denominator and the numerator makes the characteristic polynomial of the initial system, that is, the transfer function denominator $W_{cl}(s)$.

Returning to characteristic equation (1.6), one can conclude that there is some uncertainty in the choice of polynomial $H(s)$, while the polynomial $G(s)$ varies in such a manner that the characteristic polynomial $D(s)$ stays the same. Thus, the same autonomous closed-loop system can be represented by many open-loop systems. Their transfer functions (1.3) depend on where the signal loop is disconnected. The uncertainty is sometimes used to simplify the open-loop system transfer function, but in most cases the place of the disconnection is the place where the nonlinear of time-variant unit is. In this case, the linear, nonlinear, and nonstationary subsystems can be separated and it simplifies essentially the analysis. This approach is widely used hereinafter.

1.2 Amplitude–Phase–Frequency Characteristics

Definitions and properties. The output coordinate Laplace transform follows from transfer function definition (1.3):

$$X_{out}(s) = W(s)X_{in}(s). \tag{1.8}$$

Let the harmonic signal of amplitude A_{in} and frequency ω be the inputs to the system:

$$X_{in}(t) = A_{in} \sin \omega t. \tag{1.9}$$

Laplace transform of it is known as

$$X_{in}(s) = \frac{A_{in}\omega}{s^2 + \omega^2}. \tag{1.10}$$

Let us rewrite (1.8) as a sum

$$X_{out}(s) = \frac{H(s)}{G(s)} \cdot \frac{A_{in}\omega}{s^2 + \omega^2} \equiv \frac{A_{out}\omega e^{-j\psi}}{s^2 + \omega^2} + \frac{K(s)}{G(s)}. \tag{1.11}$$

Fig. 1.3 Amplitude–phase–frequency characteristic of frequency response

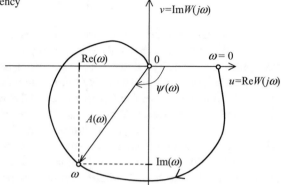

Here, the first component in the right part is the Laplace transform of the steady output coordinate harmonic oscillations at the amplitude A_{out}, the frequency ω, and the phase Ψ. The second component is the Laplace transform of the transient response, where $K(s)$ is a polynomial of $(n-1)$th degree. $A_{out}e^{j\psi_{out}}$ is obtained from the numerator balance in (1.11) taking $s = j\omega$:

$$A_{out}e^{j\psi} = \frac{H(j\omega)}{G(j\omega)}A_{in} = W(j\omega)A_{in}. \tag{1.12}$$

The function $W(j\omega)$ is called an amplitude–phase–frequency characteristic (APFC) of the system also known as frequency response. It can be depicted on the complex plane (u, v) as a hodograph, while the frequencies ω vary from 0 to ∞ (see Fig. 1.3):

$$u + jv = \mathrm{Re}W(j\omega) + j\mathrm{Im}W(j\omega) = \mathrm{Re}(\omega) + j\mathrm{Im}(\omega). \tag{1.13}$$

Here, $u = \mathrm{Re}W(j\omega) = \mathrm{Re}(\omega)$ is the real part of frequency response and $v = \mathrm{Im}W(j\omega) = \mathrm{Im}(\omega)$ is the imaginary part of frequency response. At each point ω of the hodograph, we have a vector $\vec{0\omega}$. The vector modulus $A(\omega)$ and phase $\psi(\omega)$ are

$$A(\omega) = |W(j\omega)| = \sqrt{\mathrm{Re}^2(\omega) + \mathrm{Im}^2(\omega)} \tag{1.14}$$

$$\psi(\omega) = arctg\frac{\mathrm{Im}(\omega)}{\mathrm{Re}(\omega)}. \tag{1.15}$$

Here, $A(\omega)$ is the amplitude–frequency characteristic and $\psi(\omega)$ is the phase–frequency characteristic. Amplitude–phase–frequency characteristic representation in the form of (1.13) is called complex characteristic. The frequency response can also be given in the exponential form within the amplitude (1.14) and phase (1.15) notation

$$W(j\omega) = A(\omega)e^{j\psi(\omega)}. \tag{1.16}$$

If the transfer function is strictly proper, or $n > m$, the amplitude–phase–frequency characteristic ends ($\omega \to \infty$) at the origin. In a particular case, when $n = m$, APFC ends on the real axis. The initial hodograph vector modulus ($\omega = 0$) equals to the static transfer constant of system. The origin hodograph point is on the real axis if the transfer function does not have zero poles and that point coincides with the origin of coordinates if the transfer function has zeros there, i.e., includes one (or more) operator s in the numerator as a multiplier. It is also useful to keep in mind that APFC begins at infinity along the negative imaginary semiaxis if the $W(s)$ has one zero pole and along the negative real semiaxis if the $W(s)$ has two zero poles, etc., that is, along the j^{-i}th semiaxis when there are i zero poles in the transfer function. At last, if the transfer function has pure imaginary zero poles $p = \pm j\omega_i$, the hodograph discontinues at the points ω_i, raising to infinity, while the phase jumps by the value of $-\pi$.

1.3   Dynamic System Stability and Oscillations

One of the main problems of linear dynamic system analysis is to define the stability of its unique zero equilibrium state. As is known, the necessary and sufficient condition for a linear system to be stable is that the real parts of all the roots of characteristic equation (1.6) are negative:

$$\mathrm{Re}\, s_i < 0, \quad i = 1, 2, \ldots n,$$

i.e., all the roots $s_i = \alpha_i + j\omega_i$ are to be within the left semiplane of the complex plane $(\alpha, j\omega)$. If all the characteristic equation roots are *left* ones, i.e., located in the left semiplane, except for either the zero root ($s = 0$) or a pair of imaginary (conjugated) roots ($s = \pm j\omega_0$), located on the imaginary axis, then they say that the system is at its *stability boundary*. In the first case, the stability boundary is called *aperiodic* and the system is considered *neutral*. In the second case, the stability boundary is called *oscillatory* and the system is considered *conservative*. Depending on the initial conditions, in the first case, the system can stay at any stationary point over any period of time and in the second case it can have neither convergent nor divergent oscillations of amplitude A and frequency, ω_0.

The direct stability problem solution by the determination of characteristic equation roots is not always feasible. This concerns especially high-order systems and a situation in which the mathematical model of the system is not available. In addition, as soon as a system is found to be unstable, the questions how to make it stable and how to ensure certain stability margin or oscillating index, as a rule, arise immediately. That is why there are indirect algebraic and frequency stability criteria to rigorously define the stability of a linear system without root calculations.

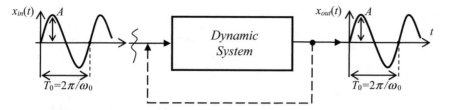

Fig. 1.4 Signal balance in feedback system

The Nyquist–Mikhaylov frequency criterion is primarily chosen from a number of known algebraic and frequency stability criterions. This criterion enables the stability analysis in a dynamic system by the system frequency characteristic obtained at the condition when one of the internal system connections is eliminated. Or, using automatic control theory terminology, the criterion permits to judge the stability of a closed system by an open-system frequency characteristic. The distinctive feature of the Nyquist–Mikhaylov frequency criterion is its simple physical interpretation. Indeed, the balance principle between phases and amplitudes of input and output harmonic oscillations of a dynamic system underlies the criterion.

Let a harmonic signal be fed to the stable dynamic system input. Let a certain frequency, ω_0, exist when the steady output and input harmonic signals have the same phases and amplitudes. In that case, the input and output steady-state signals are identical and therefore the same steady-state oscillations can be observed if the output signal is fed back as the system input, as shown in Fig. 1.4.

Then, the system becomes autonomous and will experience its natural oscillations that are identical to the forced ones. The system will be at its stability boundary. Thus, the boundary condition for oscillation stability holds if the amplitude–phase–frequency characteristic modulus equals to 1 and its phase is $-2r\pi$ ($r = 0, 1, 2, \ldots$). In other words, the amplitude–phase–frequency characteristics at the point ω_0 have to cross the point ($u = 1$, $v = 0$) of the real axis (see Fig. 1.5).

If the minus-signed output signal is fed to the system input (negative feedback), the stability oscillatory boundary condition rather changes as follows:

$$|W(j\omega_0)| = A(\omega_0) = 1, \quad \arg W(j\omega_0) = \psi(\omega_0) = \pm r\pi, \quad r = 1, 3, 5 \ldots \quad (1.17)$$

The amplitude–phase–frequency characteristic $W(j\omega)$ has to pass through the point -1 of the negative real semiaxis at the frequency ω_0 (see Fig. 1.6).

It is significant that at the same time the inverse APFC also has to pass through the point ($u_0 = -1$, $v_0 = 0$) at the frequency ω_0. The condition (1.17) follows from the phase and amplitude balance principle and guarantees one pair of imaginary roots in the closed-system characteristic equation. Since the open system is assumed to be stable, the condition (1.17) ensures for the oscillatory closed system to be conservative and capable to have steady-state oscillations with an arbitrary amplitude depending on initial conditions. Thus, a reduced Nyquist–Mikhaylov frequency criterion becomes available: the frequency characteristic $W(j\omega)$ does not have to

Fig. 1.5 Boundary stability
condition

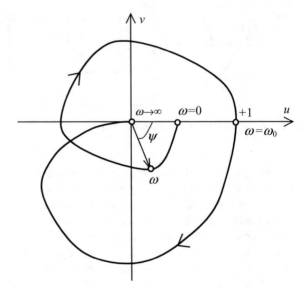

Fig. 1.6 Boundary stability
condition concerning
negative feedback system

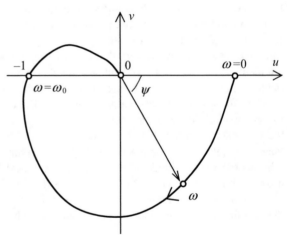

encircle the point $(-1, j0)$. Or, equivalently, a closed-loop system is steady state if
the algebraic sum of upward (negative) and downward (positive) crossings of the
real axis segment $(-\infty, -1)$ by the frequency characteristic $W(j\omega)$ is zero. If we use
inverse frequency characteristic plane, the segment $(0, -1)$ is to be used instead of
$(-\infty, -1)$ and the crossing signs are reversed. Both phase and amplitude stability
margins are derived in the amplitude–phase–frequency characteristic plane based on
the balance principle between phases and amplitudes. They focus on two important
points on APFC for that. One is *critical frequency*, ω_{cr}, at which the phase shift is
$-180°$ and another is *cutoff frequency*, ω_{cut}, at which the frequency characteristic
modulus becomes 1 (see Fig. 1.7).

Fig. 1.7 Stability margin
determination

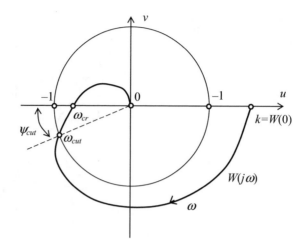

The phase stability margin is defined by the phase shift to be added to provide
the phase balance, i.e., the angle ψ_{cut}, at the cutoff frequency, ω_{cut}. The amplitude
stability margin is specified by the static gain that is needed to reach the amplitude
balance or point $(-1, j0)$ at the critical frequency. The critical transfer constant
$W(0) = k_{kr} = k/|0\omega_{kr}|$ is determined by the point ω_{kr} too. These definitions enable
the following simple formulation of closed-loop system stability condition: the cutoff
frequency is to be less than its critical frequency.

The Nyquist–Mikhaylov frequency criterion form becomes somewhat complicated concerning the general arbitrary transfer function $W(s)$. Let an unstable system have the transfer function $W(s) = H(s)/G(s)$ with **R** right poles and **I** poles
on the imaginary axis. The inverse amplitude–phase–frequency characteristic of the
closed-loop system is

$$W_{c\ell}^{-1}(j\omega) = 1 + W(j\omega) = \left.\frac{G(s) + H(s)}{G(s)}\right|_{s=j\omega} = \frac{G(j\omega) + H(j\omega)}{G(j\omega)}. \qquad (1.18)$$

Its numerator and denominator are obtained from the characteristic polynomial of
the closed- and open-loop systems, correspondingly. The phase increment for the
frequency response (1.18) in the frequency range from $-\infty$ to $+\infty$ is the following:

$$\Delta \arg W_{c\ell}^{-1}(j\omega)_{-\infty<\omega<+\infty} = \Delta \arg[G(j\omega) + H(j\omega)] - \Delta \arg G(j\omega).$$

If the closed-loop system is stable, all n nulls of the numerator of (1.18) are in the
left root semiplane and the total numerator hodograph rotation by the variation of ω
from $-\infty$ to $+\infty$ is the sum of rotations of the vectors from each zero point to the
point $j\omega$, i.e., the increment is equal to $n\pi$.

Thereby, the denominator hodograph rotation is π $(n - 2\mathbf{R} - \mathbf{I})$ (see Fig. 1.8)
and the total argument increment is

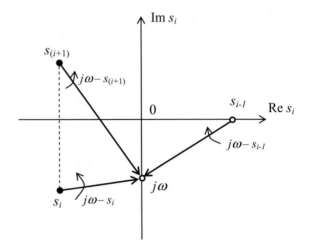

Fig. 1.8 Nyquist criterion in the case of right semiplane poles

$$\Delta \arg W_{c\ell}^{-1}(j\omega)_{-\infty<\omega<+\infty} = \pi(2R + I)$$

or by the variation of ω in the range $0 < \omega < \infty$

$$\Delta \arg W_{c\ell}^{-1}(j\omega)_{0<\omega<+\infty} = \frac{\pi}{2}(2R + I). \tag{1.19}$$

Hence, the dynamic system is stable if its inverse frequency response (1.18) rotates around the origin in positive direction by the angle $\pi(\mathbf{R} + \mathbf{I}/2)$, while the system frequency varies from zero to infinity. Here, $\mathbf{R}$ and $\mathbf{I}$ are the numbers of the unstable open-loop system transfer function poles located in the right semiplane (right poles) and on the imaginary axis, correspondingly.

As the inverse transfer function (1.18) differs from the open-loop system transfer function by one, the most popular formulation of the Nyquist–Mikhaylov frequency criterion formulation follows: a closed-loop system is stable if the rotation of the vector drawn from the point $(-1, j0)$ to the point ω of the open-loop system amplitude–phase–frequency characteristic in a range of ω from zero to infinity is positive and equal to $\pi (\mathbf{R} + \mathbf{I}/2)$. A counterclockwise rotation is considered to be positive (see Fig. 1.9).

Alternative criterion formulation requires the algebraic sum of the frequency response crossings of real axis in the interval $(-\infty, -1)$ to be positive and equal to the sum of half of total number of the right roots and a number of the imaginary root pairs of the open-loop system. The amplitude–phase–frequency characteristic of an open-loop system and its frequency response plane are called *Nyquist hodograph* and a *Nyquist hodograph plane*, correspondingly.

The frequency response function called *Mikhaylov hodograph* helps to define the stability of a dynamic system and/or a number of the characteristic equation right roots if the dynamic system is unstable. The system characteristic polynomial $D(s)$ is the Mikhaylov hodograph when s is replaced by $j\omega$. With frequency ω varying

Fig. 1.9 Nyquist–Mikhaylov criterion

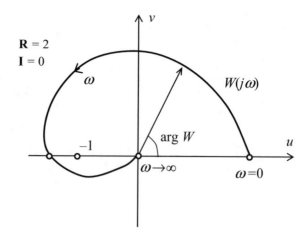

from zero to infinity, the Mikhaylov hodograph of the stable dynamic system rotates about the origin in a positive, counterclockwise, direction by the angle $n\pi/2$ visiting all n quadrants. If the condition fails, a number of the characteristic polynomial right roots is defined as a half of difference between n (system's order) and the algebraic sum of quadrants visited by the hodograph.

1.4 Examples

Let us illustrate the theory above by several simple examples. Figure 1.10 shows the Mikhaylov hodographs with regard to a third-order dynamic system. Hodograph 1 crosses the origin and represents the oscillatory stability boundary of the system. Hodograph 2 meets the stability conditions because it visits all three quadrants. Hodograph 3 and hodograph 4 represent unstable systems, the system characteristic equation has one right root in case of hodograph 3 and two roots located in the right semiplane in the case of hodograph 4.

Figure 1.11 shows the Nyquist hodographs with respect to a stable open-loop system. A closed-loop system described by hodograph 1 is stable; hodograph 3 corresponds to the unstable closed-loop system. Hodograph 2 reports that the closed-loop system belongs to the oscillatory stability boundary. As it follows from Fig. 1.11, the stability of all three systems can be reached by the change of quantitative parameters, for example, the static gain $W_i(0) = k_i$ since the frequency characteristics vary in scale in this case. This class of systems is called structurally stable. Figure 1.12 presents the inverse Nyquist hodographs of those of Fig. 1.11.

Figures 1.13, 1.14, and 1.15 show the Nyquist hodographs of unstable open-loop systems. All three systems are unstable in closed-loop architecture as well. They cannot be stabilized by quantitative parameter change. This type of systems is called structurally unstable. The only way to stabilize their closed-loop systems is to

Fig. 1.10 Mikhaylov hodographs for third-order dynamic system

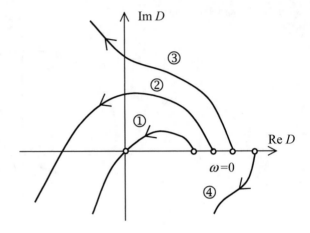

Fig. 1.11 Nyquist hodographs for stable open system

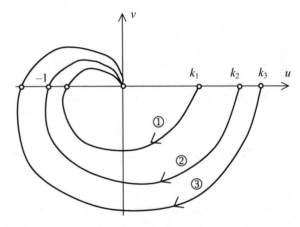

Fig. 1.12 Inverse Nyquist hodographs

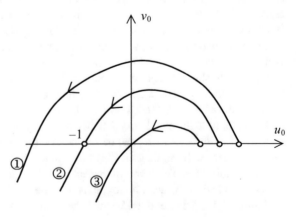

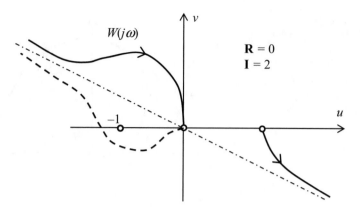

Fig. 1.13 Example 1. Unstable system hodograph

Fig. 1.14 Example 2. Unstable system hodograph

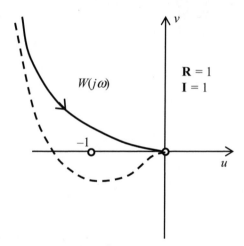

introduce the positive phase shift (phase lead correction), as it can be seen with the dotted curve segments in Figs. 1.13, 1.14, and 1.15.

As mentioned above, the frequency response can be applied to provide the qualitative and some quantitative assessments of dynamic properties. Let us consider a dynamic system described by the third-order characteristic equation in the autonomous architecture

$$T_1 T_2 T_3 s^3 + T_1 T_2 s^2 + T_1 s + 1 = 0.$$

Let the first two terms and the last two terms be the polynomial $G(s)$ and the polynomial $H(s)$, correspondingly, then the "open-loop system" transfer function takes the form

Fig. 1.15 Example 3.
Unstable system hodograph

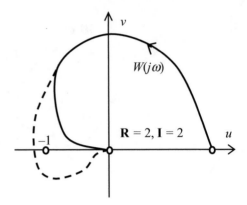

Fig. 1.16 Third-order
system stability
determination

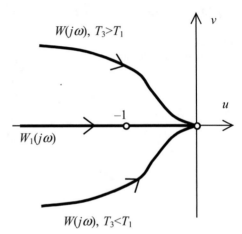

$$W(s) = \frac{1}{T_1 T_2 s^2} \cdot \frac{T_1 s + 1}{T_3 s + 1} = W_1(s) W_2(s).$$

The frequency response $W_1(j\omega)$ coincides with the negative real semiaxis (see Fig. 1.16).

Thus, the Nyquist hodograph $W(j\omega)$ belongs either the upper or lower complex semiplane depending on the sign (plus or minus) of the frequency response argument $W_2(j\omega)$. This defines the stability of the closed-loop system as long as $T_3 < T_1$. This condition is actually the necessary and sufficient stability condition for the system under consideration.

Let us turn to the fifth-order system as another example:

$$D(s) = T_1 T_2 T_3 T_4 T_5 s^5 + T_1 T_2 T_3 T_4 s^4 + T_1 T_2 T_3 s^2 + T_1 T_2 s^2 + T_1 s + 1 = 0.$$

An "open-loop system" transfer function can be written down in one of its possible forms as

Fig. 1.17 Fifth-order system stability determination

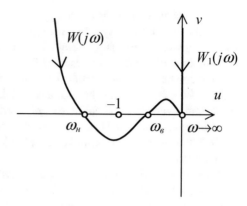

$$W(s) = \frac{1}{T_1 T_2 T_3 s^3} \cdot \frac{T_1 T_2 s^2 + T_1 s + 1}{T_4 T_5 s^2 + T_4 s + 1} = W_1(s) W_2(s).$$

The frequency response of the first factor coincides with the positive imaginary semi axis (see Fig. 1.17).

If the open-loop frequency response is supposed to encircle the point $(-1, j0)$ in the positive direction (or in other words, if a stability domain is supposed to exist), the argument of $W_2(j\omega)$ is required to be greater than $+\pi/2$. In turn, for doing so the nominator's resonance frequency $\omega_7 = 1/T_7 = 1/\sqrt{T_4 T_5}$ needs to be less than that of denominator, i.e., $\omega_6 = 1/T_6 = 1/\sqrt{T_1 T_2}$ and also the phase of the frequency response $W_2(j\omega)$ is to be about $+\pi/2$ at those resonance frequencies. For that the constants T_2 и T_4 have to be sufficiently apart from each other. In practice, the search of system stability domain can be based on simple approximate conditions

$$\omega_7 > 10\omega_6, \quad T_2 > 10 T_7, \quad T_6 < 10 T_4.$$

In this case by varying the time constant T_3, or the static gain, one can achieve for the point $(-1, j0)$ to be in a range of (ω_s, ω_b), as it is shown in Fig. 1.17.

1.5 Distributed Parameter Systems

Some technological processes are distributed in time and in space. The propagation of heat fluxes in a furnace volume and the oscillations in elastic and liquid media and the transmission of signals and energy through long electric lines and networks are dependent not only on time but also on measurement point locations.

The well-known illustration of a distributed process is the vibration of a string fixed at its both ends. The oscillation amplitude depends not only on time but the distance between the string ends.

The propagation of a signal through a long electrical line should also be counted as distributed parameter system framework. In mechanics, the distributed parameters such as mass, stiffness, damping, and inertia moment are widely used to describe the continuum media and can be essential for understanding the oscillatory behavior.

Mathematics employs partial derivatives to model the distributed systems. In general, finding the exact solutions of partial derivative equations is complicated and successful for countable simple cases only. Fortunately, in most situations, exact solutions are not required and can be replaced by the frequency responses (see Chap. 2).

Both the operator and complex amplitude methods are illustrated below with classic examples.

Long electrical line. Let us consider a semi-infinite current line without ohmic power losses, which is characterized by the distributed inductance L and capacity C. The electromotive force $e(t)$ is applied at the left end of the line ($x = 0$). The internal line processes are described by the differential equation

$$\frac{\partial^2 u}{\partial t^2} = a^2 \frac{\partial^2 u}{\partial x^2}, \quad a = \frac{1}{\sqrt{LC}}$$

with the initial and boundary conditions

$$u(x, 0) = 0; \quad \frac{\partial u(x, 0)}{\partial t} = 0; \quad u(0, t) = e(t).$$

The equation has the operator form

$$s^2 U(x, s) = a^2 \frac{\partial^2 U(x, s)}{\partial x^2}$$

and the boundary condition becomes

$$U(0, s) = E(s).$$

The solution is given in the form

$$U(x, s) = C_1 e^{-\frac{sx}{a}} + C_2 e^{\frac{sx}{a}}.$$

As $U(x, s)$ is a bounded function at $x \to \infty$, a constant C_2 has been equal to zero. The constant, C_1, is obtained from the boundary condition as follows:

$$U(0, s) = E(s) = C_1.$$

Hence

$$U(x, s) = E(s)e^{-\frac{sx}{a}}.$$

It means that an ideal distributed line has the transport lag $\tau = x/a$. The lag value is proportional to the coordinate, x, and depends on the wave velocity, a. The frequency response is defined by substituting s by $j\omega$.

Temperature distribution. A thin homogeneous infinite heat-insulated beam is heated at its left end ($x = 0$) under the impact of function $u(0, t) = f(t)$. The initial temperature distribution is zero, and then its distribution is governed by the equation

$$\frac{\partial u}{\partial t} = a^2 \frac{\partial^2 u}{\partial x^2},$$

and the boundary conditions

$$u(x, 0) = u_0; \quad u(0, t) = f(t).$$

The operator form of the equation is

$$sU(x, s) = a^2 \frac{\partial^2 U(x, s)}{\partial x^2}$$

under the boundary condition

$$U(0, s) = F(s).$$

The equation has a solution

$$U(x, s) = C_1 e^{-\frac{x\sqrt{s}}{a}} + C_2 e^{\frac{x\sqrt{s}}{a}}.$$

Here, $C_2 = 0$ due to the physical sense and $C_1 = F(s)$ by the boundary condition. Thus

$$U(x, s) = F(s)e^{-\frac{x\sqrt{s}}{a}}.$$

Chapter 2
Discrete Systems

The chapter deals with discrete-time processes and discrete modeling.

2.1 Discrete-Time Processes

Whereas continuous processes are described by ordinary differential equations, discrete processes are described by equations in differences. Here is one of the difference equation examples:

$$f(n) + f(n-1) + f(n-2) + \cdots + f(n-m) = 0,$$

where $n = 1, 2, \ldots$ is the number of time range period, T, and $t = nT$ is discrete time, $m \leq n$.

Similar to continuous systems, the difference equation has its operator form, which is given through z-transform as

$$F(z) = \sum_{n=0}^{\infty} f(n)z^{-n}, \quad z = e^{pT}$$

denoted as $F(z) = Z[f(z)]$. So the operator form of the difference equation is

$$(1 + z^{-1} + z^{-2} + \cdots + z^{-m})F(z) = 0.$$

Example. Let the discrete exponential function

$$f(n) = e^{-\gamma Tn}, \quad \gamma = \alpha + j\beta$$

have the z-transformation

© Springer International Publishing AG, part of Springer Nature 2017
L. Chechurin and S. Chechurin, *Physical Fundamentals of Oscillations*,
https://doi.org/10.1007/978-3-319-75154-2_2

Table 2.1 z-transforms and operator forms for the continuous systems

$f(t)$	$F(s)$	$f(n)$	$F(z)$	$F(z, \varepsilon)$
$\delta(t)$	1	$\delta(n)$	1	1
$1(t)$	$\dfrac{1}{p}$	$1(n)$	$\dfrac{z}{z-1}$	$\dfrac{z}{z-1}$
t	$\dfrac{1}{p^2}$	nT	$\dfrac{Tz}{(z-1)^2}$	$\dfrac{Tz}{(z-1)^2} + \dfrac{\varepsilon T z}{z-1}$
$e^{-\alpha t}$	$\dfrac{1}{p+\alpha}$	$e^{-\alpha n t}$	$\dfrac{z}{z-d}, d = e^{-\alpha T}$	$\dfrac{z}{z-d} d^{\varepsilon}$
$e^{-\alpha t} \sin \beta t$	$\dfrac{\beta}{(p+\alpha)^2 + \beta^2}$	$e^{-\alpha n T} \sin \beta n T$	$\dfrac{zd \sin \beta T}{z^2 - 2zd \cos \beta T + d^2}$	$zd^{\varepsilon} \dfrac{z \sin \varepsilon \beta T + d \sin(1-\varepsilon)\beta T}{z^2 - 2zd \cos \beta T + d^2}$
$e^{-\alpha t} \cos \beta t$	$\dfrac{p+\alpha}{(p+\alpha)^2 + \beta^2}$	$e^{-\alpha n T} \cos \beta n T$	$\dfrac{z^2 - zd \cos \beta T}{z^2 - 2zd \cos \beta T + d^2}$	$zd^{\varepsilon} \dfrac{z \cos \varepsilon \beta T - d \cos(1-\varepsilon)\beta T}{z^2 - 2zd \cos \beta T + d^2}$

$$F(z) = \sum_{n=0}^{\infty} e^{-\gamma T n} z^{-n} = d \sum_{n=0}^{\infty} \left(\frac{z}{d}\right)^{-n} = \frac{z}{z-d}, d = e^{-\alpha T}, \beta = 0.$$

For complex γ, the initial function takes the form

$$f(n) = e^{-\alpha T n}(\cos \beta T n - j \sin \beta T n).$$

By separating the real and imaginary parts, $F(z)$ is written down as

$$F(z) = \frac{z^2 - zd \cos \beta T - jzd \sin \beta T}{z^2 + 2dz \cos \beta T + d^2}.$$

Table 2.1 follows from the last relation for z-transforms with the addition of operator forms for the continuous systems.

The substitution $z = e^{j\omega T}$ gives the transfer from the operator form to the frequency characteristic. The z-transform operator is a periodical function, and the frequency characteristics of the discrete-time systems are also periodical with a period of $\omega = \pi/T$.

2.2 Discrete Parameter Models

Several connected identical pendulums or the train of several carriages connected with each other by buffering are examples of chained systems. They are discrete systems which are also described by difference equations. Whereas the discrete-time argument, n, is the number of time periods, that concerning chained system is the unit's number.

There is another reason to apply a chain system description. As mentioned above (see p. 5, Chap. 1), the exact solutions of partial derivative equations have a complicated and rather exclusive character. That is why the problem of modeling has

Fig. 2.1 Chain circuit

a great importance with regard to distributed systems. The chained description for the approximation of distributed systems can be called a discrete modeling which is similar to a well-known finite element method (FEM). As to the chain systems, discrete modeling gives exact solutions.

Illustration of method. Let us consider a simple example. Figure 2.1 presents the chain dynamic system consisting of the identical sequential elements $W(s)$.

The difference equation in the denotations of input $x(n)$ and output $x(n + 1)$ is as follows:

$$x(n + 1) = W(s)x(n).$$

Using the z-transformation, the last equation takes the form

$$zX(z) - zx(0) = W(s)X(z).$$

Hence

$$X(z) = \frac{z}{z - W(s)}x(0).$$

The following result is derived as a result of the inverse z-transformation (see Table 2.1):

$$x(n, s) = [W(s)]^n x(0).$$

It is a quite apparent result since just the illustrative system has no intersecting connections.

So, the cardinal method consists of dividing a system into identical finite elements, generating difference-differential equations for a separated element, and solving the equations based on both direct and inverse z-transforms. In this way, the frequency characteristics of distributed systems can be defined.

Modeling of long electrical line. Let us consider a modeling method by the example of a long electrical line. Let $z_1(0)$ denote either of two sequential resistances and $z_2(0)$ denote a parallel resistance, all are adopted per a unit line length. Dividing the line into simple elements, the nth element constitutes the electrical circuit shown in Fig. 2.2.

The element obeys two the difference equations:

$$\begin{aligned} (z_1 + z_2)\,i(n) - z_2 i(n + 1) - u(n) &= 0 \\ z_2 i(n) - (z_1 + z_2)\,i(n + 1) - u(n + 1) &= 0 \end{aligned} \tag{2.1}$$

Fig. 2.2 Discrete model of long electrical line

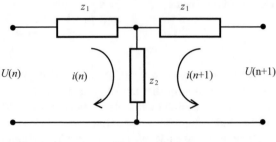

Fig. 2.3 Structural representation of long electrical line

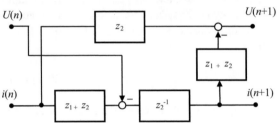

Figure 2.3 provides the structural representation of the single element. The elements compose the chained system.

Specifying the left boundary conditions as $u(0)$ and $i(0)$ and resorting to the z-transformation, system (2.1) takes the form

$$(z_1 + z_2 - z_2 z)I(z) - U(z) = -z_2 z i(0)$$
$$[z_2 - (z_1 + z_2) z]I(z) - zU(z) = - (z_1 + z_2) z i(0) - zu(0). \tag{2.2}$$

This algebraic system has its solution as to the z-transformations of $I(z)$ and $U(z)$:

$$I(z) = \frac{z(z - ch\chi) i(0) - z z_2^{-1} u(0)}{z^2 - 2z ch\chi + 1}$$
$$U(z) = \frac{z(z - ch\chi) u(0) - (z z_2 sh^2 \chi) i(0)}{z^2 - 2z ch\chi + 1}, \tag{2.3}$$

where $ch\chi = 1 + z_1 z_2^{-1}$ is hyperbolic cosine.

The use of the inverse z-transform (see Table 2.1) gives

$$i(n) = i(0)ch\chi n - \frac{u(0)}{z_2 sh\chi} ch\chi n$$
$$u(n) = u(0)ch\chi n - i(0)z_2 sh\chi \, sh\chi n. \tag{2.4}$$

Let $R(s)$ and $Y(s)$ denote operator resistance and conductance, correspondingly, then

$$R(s) = \frac{1}{Y(s)} = z_2 sh\chi = z_2 \left[\frac{z_1}{z} \left(2 + \frac{z_1}{z_2} \right) \right]^{1/2}$$

and solution (2.4) takes the form

$$i(n) = i(0)ch\chi n - u(0)Y(p)sh\chi n \tag{2.5}$$

$$u(n) = u(0)ch\chi n - i(0)R(p)sh\chi n. \tag{2.6}$$

By virtue of interactions of all the elements, the left boundary conditions are related to those for the right boundary when $n = N$. If the line is short-circuited at the end, that is $u(N) = 0$, the relation between the conditions is obtained from (2.6) as

$$i(0) = u(0)\frac{ch\chi N}{R(p)sh\chi n} = u(0)Y(p)cth\chi n$$

and solutions (2.5) and (2.6) can be expressed as follows:

$$i_c(n, p) = u(0)Y(p)\,(cth\chi Nch\chi n - sh\chi n) = u(0)Y(p)\frac{ch\chi\,(N-n)}{sh\chi N} \tag{2.7}$$

$$u_c(n, p) = u(0)(ch\chi n - cth\chi Nsh\chi n) = u(0)\frac{sh\chi\,(N-n)}{sh\chi N}. \tag{2.8}$$

If the N th element is open, $i(N) = 0$, solution (2.5) gives

$$i(0) = u(0)Y(p)\frac{sh\chi N}{ch\chi N} = u(0)Y(p)th\chi N.$$

In this case, the solutions follow from (2.5) to (2.6) as

$$i_o(n, p) = u(0)Y(p)\,(th\chi nch\chi n - sh\chi n) = u(0)Y(p)\frac{sh\chi\,(N-n)}{ch\chi N} \tag{2.9}$$

$$u_o(n, p) = u(0)\,(ch\chi n - th\chi Nsh\chi n) = u(0)\frac{ch\chi\,(N-n)}{ch\chi N}. \tag{2.10}$$

Expressions (2.7) to (2.10) provide the transfer functions and the frequency characteristics as long as $s = j\omega$ at any chosen point of the electrical line. For example, from (2.8) to (2.10) the frequency characteristics are as follows:

$$W_c(j\omega, n) = \frac{u_c(n, j\omega)}{u(0)} = \frac{sh[(N-n)\chi(j\omega)]}{sh[N\chi(j\omega)]}$$

$$W_o(j\omega, n) = \frac{u_o(n, j\omega)}{u(0)} = \frac{ch[(N-n)\chi(j\omega)]}{ch[N\chi(j\omega)]}.$$

The frequency characteristics are well known to describe main system properties such as stability, oscillateness, transient responses, etc. They are used in Part 5 hereinafter.

Chapter 3
Experimental and Numerical Evaluations of Frequency Response

3.1 Experimental Frequency Response Evaluation

With exponential form of the amplitude–phase–frequency characteristic (1.16), relationship (1.12) takes the form

$$A_{out} e^{-j\psi\ out} = A(\omega) e^{-j\psi(\omega)} A_{in}. \tag{3.1}$$

Equating modules and phases from the sides of formulation (3.1), the following relations are obtained:

$$\begin{aligned} A_{out}/A_{in} &= A(\omega) \\ \psi_{out} &= \psi(\omega). \end{aligned} \tag{3.2}$$

Thus, the amplitude–frequency characteristic or frequency response at the point ω, is the ratio of the steady-state harmonic oscillation amplitude at the system output to the harmonic oscillation amplitude with frequency ω at the system input; the phase-frequency characteristic is the phase–frequency dependence regarding the output steady-state harmonic oscillations.

There are several ways to obtain the system's frequency response from physical experiment [3]. The simplest approach based on formulation (3.2) is to feed an input harmonic wave with amplitude, A_{in}, and frequency, ω. Then, both the amplitude and the phase of the output steady-state oscillation are measured.

The point ω of APFC is plotted in the complex plane (Fig. 1.3). The point has modulus equal to the amplitude ratio, and its phase is that of the output oscillations. The above procedure is reiterated using different oscillation frequencies. Hence, the experimental frequency characterization plant includes the sinusoidal generator *Generator* and the double-beam recorder *Scope* (see Fig. 3.1). A personal computer with a DA/AD converter interface card is used to characterize and process of the frequency characteristics.

© Springer International Publishing AG, part of Springer Nature 2017
L. Chechurin and S. Chechurin, *Physical Fundamentals of Oscillations*,
https://doi.org/10.1007/978-3-319-75154-2_3

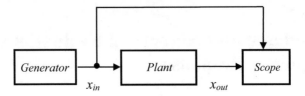

Fig. 3.1 Plant scheme for experimental APFC determination

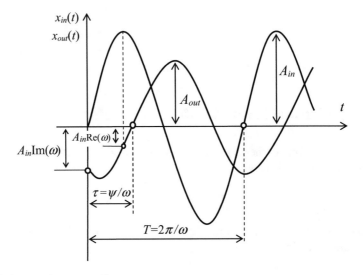

Fig. 3.2 Input and output oscillograms

Figure 3.2 shows the signals of the input ($x_{in}(t) = A_{in} \sin \omega t$) and output ($x_{out}(t) = A_{out} \sin (\omega t - \psi)$) as steady-state oscillations. Apart from the amplitude–and phase–frequency characteristics, the oscilloscope signal depicts the values of real ($\text{Re}(\omega) = A(\omega) \cos \psi$) and imaginary ($\text{Im}(\omega) = A(\omega)\sin \psi$) frequency characteristics.

3.2 Numerical Evaluation of Frequency Responses

The direct evaluation of amplitude–phase–frequency characteristic includes (a) the separation of real and imaginary parts in numerator and denominator of the transfer function $W(s)$ at $s = j\omega$, and (b) the multiplication of the numerator and the divisor by the conjugate divisor as follows:

$$W(j\omega) = \frac{H(j\omega)}{G(j\omega)} = \frac{\text{Re }H + j\text{Im}H}{\text{Re}G + j\,\text{Im }G} \cdot \frac{\text{Re }G - j\,\text{Im }G}{\text{Re}G - j\,\text{Im }G}$$

$$= \frac{(\text{Re}H\,\text{Re}G + \text{Im}H\,\text{Im}G) + j(\text{Im}H\,\text{Re}G - \text{Re}H\,\text{Im}G)}{\text{Re}^2 G + \text{Im}^2 H}$$

$$= \text{Re}W(j\omega) + j\text{Im }W(j\omega).$$

Another method is the calculation of the modulus and the phase according to formulations (1.14) to (1.15). The simpler approach is the calculation of the modulus and the phase of the APFC numerator and denominator

$$W(j\omega) = \frac{|H(j\omega)|\,e^{j\psi_H(\omega)}}{|G(j\omega)|\,e^{j\psi_G(\omega)}} = \frac{\sqrt{\text{Re}^2 H + \text{Im}^2 H}}{\sqrt{\text{Re}^2 G + \text{Im}^2 G}}e^{j[\psi_H(\omega) - \psi_G(\omega)]} = A(\omega)e^{j\psi(\omega)}.$$

When the transfer function numerator is of a zero-order or a much less order than the denominator, it is better to use the inverse amplitude–phase–frequency characteristic

$$W^{-1}(j\omega) = \frac{G(j\omega)}{H(j\omega)} = A^{-1}(\omega)e^{-\psi(\omega)} \tag{3.3}$$

and plot the inverse hodograph in the plane of the inverse amplitude–phase–frequency characteristic $[u_0 = \text{Re}W^{-1}(j\omega), v_0 = \text{Im}W^{-1}(j\omega)]$ (see Fig. 3.3).

The point ω is transferred from the direct plane to the inverse one and vice versa by replacing a modulus with an inverse one and a phase with an opposite in sign.

As a rule, the frequency characteristics $W(j\omega)$ are experimentally measured by means of feeding the input harmonic and subsequently measuring the output signal at the connection discontinuity of the system. In so doing, the "closed" system frequency characteristic is retrieved using relation (1.5) by substituting s by $j\omega$. That process is simplified in the inverse frequency characteristic plane because

$$W_c^{-1}(j\omega) = W^{-1}(j\omega) + 1, \tag{3.4}$$

Fig. 3.3 Inverse APFCs

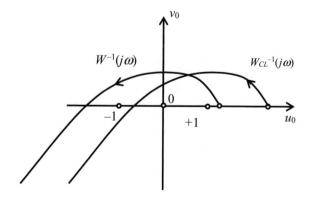

i.e., reduced to a unit displacement of the inverse amplitude–phase–frequency characteristic $W^{-1}(j\omega)$ to the right along the real axis (see Fig. 3.3).

A dynamic system is quite often occurred to become unstable as a result of any connection discontinuity, and the measurements of the frequency characteristics of the similar "open" systems are not possible. In that situation and also when an initial system is unstable, a stabilizing connection is introduced and then the closed system amplitude–phase characteristic is measured. The input harmonic is summed with the chosen connection signal, and output oscillations are measured either upstream or downstream the adder. In the first case, the experimental frequency characteristic is described by formulation (1.5) with $s = j\omega$ and the frequency characteristic of the initial "open" system is computed using formulation (1.7) or the inverse frequency characteristic $W^{-1}(j\omega)$ is plotted according to relationship (3.4) to shift the experimental inverse frequency characteristic by a unit to the left along the real axis. If the output oscillations are measured at the adder output, the experimental and calculated frequency characteristics are as follows:

$$W_{c\ell}(j\omega) = \frac{1}{1 + W(j\omega)}, \quad W(j\omega) = \frac{1 - W_{c\ell}(j\omega)}{W_{c\ell}(j\omega)} = W_{c\ell}^{-1}(\omega) - 1. \quad (3.5)$$

In conclusion, it should be noted that at present the plotting of computer amplitude–phase–frequency characteristics for high-order dynamic systems is easily accessible and not difficult. At the same time, a researcher who is able to use the available frequency characteristic apparatus can easily obtain qualitative results on dynamic system property assessments.

Part II
Parametric Oscillations of Linear Periodically Nonstationary Systems

We use one-frequency harmonic approximation to define the conditions of parametric oscillation excitement in linear periodically time-variant dynamic systems. We analyze then periodically time-variant systems with multifrequency stationary part and distributed parameters.

Chapter 4
The First Parametric Resonance

Linear periodically nonstationary systems are described by linear differential equations with either one or several periodically variable coefficients. In contrast to the coordinates, that describe the motion of the system, the periodically variable coefficients are called periodically variable parameters. The system coordinate oscillations concerned with the cyclic variations of parameters are called parametric oscillations. The parametric oscillations which occur in a linear dynamic system can rise indefinitely in their amplitude. This phenomenon is called a parametric resonance.

Not all of the dynamic system oscillations caused by parameter variations lead to the parametric resonance. So, if the equilibrium of the system is changed as a result of the variation of system parameters, certain steady-state equilibrium oscillations will correspond to the cyclic parameter variation. This takes place, for example, in the case of a spring-suspended load while the slow cycling of spring stiffness happens, and also in an electric circuit provided with an alternating resistance voltage source. The steady load displacements and the current changes in the above examples both take place at all the frequencies and amplitudes of the parameter variation. The oscillations should be called forced oscillations of the linear dynamic system. And just under defined conditions and within specified ranges of the parameter variation frequencies and amplitudes, the parametric resonance can emerge in the form of indefinitely rising oscillations.

Another reason for the oscillations in the linear dynamic system with the periodic parameter variation is the periodic variation of its single stability equilibrium. These oscillations are not the parametric resonance; they emerge at any of the parameter variation frequencies and also demonstrate the unlimited growth. The oscillations should be rather called *natural* because they are related to the variation of the system's own status. The parametric resonance oscillations can interfere with the above oscillations under certain conditions and both change the stable equilibrium of the system into unstable one and, on the contrary, stabilize the initial unstable system equilibrium.

Thus, the parametric resonance phenomenon and its excitation conditions in the dynamic systems are the main objects of study of this chapter.

© Springer International Publishing AG, part of Springer Nature 2017   33
L. Chechurin and S. Chechurin, *Physical Fundamentals of Oscillations*,
https://doi.org/10.1007/978-3-319-75154-2_4

The approach of the frequency analysis presented in this chapter was first published by the author in 1979 [4] and developed in the number of consequent papers [5–8], collected in the book [9].

4.1   Basic Frequency Relations

Let us consider an elementary periodic single-frequency parameter

$$a(t - \tau) = a \sin \Omega(t - \tau), \quad \Omega = 2\pi/T, \tag{4.1}$$

where τ reflects some arbitrary time shift or lag between the periodic parameter profile and the periodic input signal $x(t)$ at frequency ω, (see Fig. 4.1):

$$x(t) = x_0 + \tilde{x}(t) = x_0 + A \sin \omega t. \tag{4.2}$$

The output signal of the periodic element has the form

$$y(t) = a(t - \tau)x(t) = ax_0 \sin \Omega(t - \tau) + Aa \sin \Omega(t - \tau) \sin \omega t \tag{4.3}$$

and includes the following components:

$$y(t) = ax_0 \sin(\Omega t - \varphi) + \frac{Aa}{2} \cos[(\Omega - \omega)t - \varphi] + \frac{Aa}{2} \cos[(\Omega + \omega)t - \varphi] \tag{4.4}$$

of the frequency Ω, the difference $(\Omega - \omega)$ and sum $(\Omega + \omega)$ frequencies. Here the time shift τ, is replaced by the phase shift $\varphi = \Omega\tau$.

To say that the periodically nonstationary element transmits the input signal of frequency ω, (see Fig. 4.1), we need to observe at least one of the components of the same frequency among the output signal components. Having assumed all the frequencies positive, which is a real physical requirement, we arrive at two conditions

$$\omega = \frac{\Omega}{2}, \tag{4.5}$$

$$\omega = \Omega. \tag{4.6}$$

Fig. 4.1 Periodic parameter unit

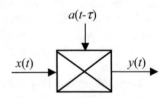

When the condition (4.5) is satisfied, the sustained parametric oscillations $x(t)$ emerge. They are called *the first parametric resonance*. The parametric oscillations at the parameter variation frequency of condition (4.6) are called *second parametric resonance* oscillations. The conditions (4.5) and (4.6) tell that the first parametric resonance oscillations are $2T$-periodic and the second ones are T-periodic. The first and second parametric resonances are basic, they completely describe the single-frequency oscillations of the autonomous linear periodically time-variant systems. Let us consider them separately hereinafter.

4.2 The First Parametric Resonance

Transfer function of a periodic parameter. Let the periodic parameter vary according to the following profile:

$$a(t) = a_0 + a \sin \Omega(t - \tau), \quad \Omega = 2\pi/T, \tag{4.7}$$

whereas the input signal has the form

$$x(t) = \tilde{x}(t) = A \sin \frac{\Omega}{2} t. \tag{4.8}$$

In this case, the output signal of the periodic parameter is

$$y(t) = a(t)x(t) = Aa_0 \sin \frac{\Omega}{2} t + Aa \sin \Omega(t - \tau) \sin \frac{\Omega}{2} t.$$

Having decomposed the product of sinuses into the sum of ones, we distinguish the terms with input frequencies $\omega = \Omega/$in the output signal as follows:

$$\tilde{y}(t) = Aa_0 \sin \frac{\Omega}{2} t + \frac{Aa}{2} \cos \frac{\Omega}{2}(t - 2\tau) = Aa_0 \sin \frac{\Omega}{2} t + p \frac{a}{\Omega} A \sin \frac{\Omega}{2}(t - 2\tau), \tag{4.9}$$

where s is an operator of differentiation. Taking into account input signal (4.8), the expression (4.9) becomes

$$\tilde{y}(t) = a_0 \tilde{x}(t) + \frac{a}{\Omega} s\tilde{x}(t - 2\tau). \tag{4.10}$$

Since $\tilde{y}, \tilde{x}$ are harmonic functions, the equality (4.10) can be given the symbolic complex form. We note that $s = j\Omega/2$ and the shift in the time domain corresponds to multiplication by $\exp(-j\Omega\tau)$

$$\tilde{Y}(j\Omega/2) = a_0 \tilde{X}(j\Omega/2) + \frac{ja}{2} e^{-j\Omega\tau} \tilde{X}(j\Omega/2). \tag{4.11}$$

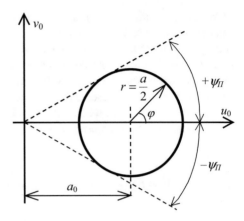

Fig. 4.2 First parametric resonance circle

Let us compose the following ratio, that is, the frequency response for the transfer function of the periodic parameter:

$$\frac{\tilde{Y}(j\Omega/2)}{\tilde{X}(j\Omega/2)} = W(j\varphi) = a_0 + \frac{ja}{2}e^{-j\varphi}, \quad \varphi = \Omega\tau. \qquad (4.12)$$

Thus, the transfer function of the periodically nonstationary element takes its final form

$$W(j\varphi) = a_0 - \frac{a}{2j}e^{-j\varphi} = a_0 - \rho e^{-j\varphi}, \quad \rho = a/2j. \qquad (4.13)$$

The form (4.13) was obtained in cooperation with Ostrovskii [6].

Let us point out several important clarifications here.

1. In the complex plane [$u_0 = \mathrm{Re}W(j\varphi)$, $v_0 = \mathrm{Im}W(j\varphi)$], the frequency response function (4.13) is the circle of radius $r = |\rho| = |a/2j| = a/2$ centered at the point a_0, (see Fig. 4.2).
2. The transfer function (4.13) was derived only for the input signal of certain frequency, that is double-parameter variation frequency. It means that it is legitimate model of the periodic parameter element for the first parametric resonance only.
3. The derived transfer function does not depend on the input amplitude A. It reflects the fact that periodic parameter is a linear element.
4. The derived transfer function is a new type of description, since it does not depend on the input amplitude and depends essentially on the phase shift between the input signal and the periodic parameter.
5. The transfer function (4.13) implies that the periodic parameter can add to the input signal an additional extra positive as well as negative phase shift. So it is either the phase lead ($+\psi_n$) or the phase lag ($-\psi_n$) (see Fig. 4.2). To explain a phase shift effect, let us examine an amplifier as an elementary periodically nonstationary element. Let the amplifier gain jumps from 0 to 2, as it shown

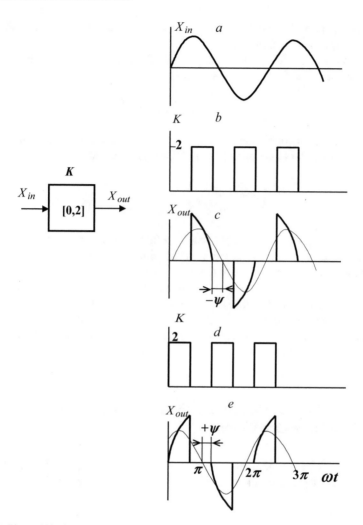

Fig. 4.3 Phase shift phenomenon illustration

in Fig. 4.3b. The output signal in Fig. 4.3c corresponds to the input harmonic in Fig. 4.3a. The output signal has many harmonics and its first harmonic is separated (see a dotted line), the phase shift $+\psi$ is revealed by comparing the output first harmonic component with the input signal. The similar conclusion arises in the other relative position between the input signal and the parameter (see Fig. 4.3d). In that case the phase advance to the first harmonic of the output signal is fixed.

6. The initial parametric element is nonstationary and depends on time, whereas description (4.13) is steady-state and does not depend on time. This allows the

replacement of the nonstationary parameter by the description (4.13). This procedure is called *stationarization*.

7. The transfer function (4.13) is described on (u_0, v_0) plane of the inversed frequency response function as the circle

$$(u_0 - a_0)^2 + v_0^2 = r^2. \tag{4.14}$$

On the complex coordinate plane $[u = \mathrm{Re}W^{-1}(j\varphi), v = \mathrm{Im}W^{-1}(j\varphi)]$ the inversed transfer function of the parameter $W^{-1}(j\varphi)$ is also a circle

$$\left(u - \frac{a_0}{a_0^2 - r^2}\right)^2 + v^2 = \frac{r^2}{\left(a_0^2 - r^2\right)^2}, \tag{4.15}$$

where $a_0/(a_0^2 - r^2)$ is a center and $r_0 = r/(a_0^2 - r^2)$ is a radius.

8. In the case of symmetric variation of the parameter, the constant component of the parameter variation is zero ($a_0 = 0$) and transfer function (4.13) is a central circle of radius r. If the constant component of the parameter variation exists, it is more convenient to delegate it to the stationary part of the system. That is what the we do through all the book further. In this case the parametric transfer functions $W(j\varphi)$ and its inverse $W^{-1}(j\varphi)$ are the circles of radius $a/2$ and $2/a$ and center at the origin.

4.3   Excitation Conditions for the First Parametric Resonance Oscillation

Let us consider the autonomous dynamic system which consists of a linear stationary subsystem and the T-periodic element $a(t)$. The system is described in the operator form

$$G(s)x(t) + H(s)[a(t)x(t)] = 0 \tag{4.16}$$

and has the structure representation shown in Fig. 4.4 where the linear stationary subsystem is the transfer function $W(s) = H(s)/G(s)$.

Let the stationary subsystem is a low-frequency or resonance filter and there is only one harmonic with a frequency $\omega = \Omega/2$, i.e., first parametric resonance oscillation, at the stationary subsystem output, the approximate solution of Eq. (4.16) can be obtained by replacing the parameter $a(t - \tau)$ by transfer function (4.13) and transferring to the frequency equation using the substitution $s = j\omega = j\Omega/2$:

$$G\left(j\frac{\Omega}{2}\right) + H\left(j\frac{\Omega}{2}\right)W(j\varphi) = 0 \tag{4.17}$$

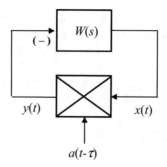

Fig. 4.4 Structure representation of periodically nonstationary system

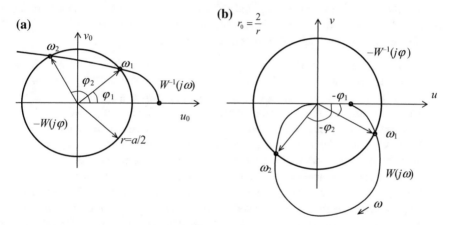

Fig. 4.5 First parametric resonance excitation conditions

Expressing the frequency characteristic of the stationary subsystem as $W(j\omega) = H(j\omega)/G(j\omega)$,

$$1 + W\left(j\frac{\Omega}{2}\right)W(j\varphi) = 0,$$

$$W\left(j\frac{\Omega}{2}\right) = -W^{-1}(j\varphi), \tag{4.18}$$

$$W^{-1}\left(j\frac{\Omega}{2}\right) = -W(j\varphi). \tag{4.19}$$

This is the first parametric resonance excitation condition and the statement follows that in the T-periodic nonstationary dynamic system the $2T$-periodic oscillations of the first parametric resonance are excited as soon as the point $\omega = \Omega/2$ of the inverse Nyquist hodograph $W^{-1}(j\omega) = G(j\omega)/R(j\omega)$ gets to the first parametric resonance circle. Figure 4.5a illustrates the above definition.

Based on condition (4.18) we can formulate now the following statement for the Nyquist plane. The periodically nonstationary system experiences the parametric resonance if the point $\omega = \Omega/2$ of Nyquist hodograph lies outside of the inversed parametric resonance circle (Fig. 4.5b).

Conditions (4.18) and (4.19) can deliver the equation for the excitation boundaries in the modulus form in both cases

$$\left| W^{-1}\left(j\frac{\Omega}{2}\right) \right| = |W(j\varphi)| = \frac{a}{2} \quad \text{or} \quad \left| W\left(j\frac{\Omega}{2}\right) \right| = |W^{-1}(j\varphi)| = \frac{2}{a}. \quad (4.20)$$

The excitation area boundary to the first parametric resonance can be evaluated by formula (4.20) in the coordinates (Ω, a), here Ω is the parameter variation frequency and a is the parameter variation amplitude. The numerical examples with the systems described by second-order differential equations are presented hereafter.

We have to give here several important clarifications.

1. There is a simple physical explanation of the excitation of the parametric resonance. It was noted that a parametric element can cause a lead/lag shift in the input signal phase depending on the phase, φ, between the oscillations of the parameter and the input parameter coordinate. In other words, the parametric element can introduce an extra positive/negative phase shift into the closed-loop of the system. If the introduced phase shift is either equal or exceeding the phase stability margin of the stationary system, the parametric resonance is excited in the periodically nonstationary system. In turn, the occurrence of the parametric resonance in the system means the loss of system stability. The excitation of parametric resonance is followed by the energy transfer from the parameter variation to the coordinate oscillations. As the magnitude of parameter variation is constant, the amount of energy injection is the constant too. It means that the growth of coordinate oscillation is linear.

2. If the point $\omega = \Omega/2$ of the inversed frequency hodograph is outside of the parametric resonance circle of radius $a/2$, there is no parametric resonance excitation. In contrast to linear time-invariant systems, there are the parametric resonance boundaries in time-invariant case. They depend on the frequency and the magnitude of the parameter a. Therefore, the excitation depends on the initial conditions too, that tell how far is the system from the zero equilibrium state.

3. The (lower) boundary for the parametric resonance excitation can be defined by the minimal distance from the point $\omega = \Omega/2$ of the inversed frequency hodograph to the parametric circle in respect to the radius

$$h = \frac{W^{-1}(j\omega) - r}{r}.$$

Most systems studied in the book are highly oscillatory and the excitation boundary is much less than 1. In other words, the excitation of the parametric resonance takes place at already small magnitudes of the parameter variation. The most interesting question is qualitative: whether the excitement happens or not when the variations are small. It legitimates the main idea of the approach, when we approximately evaluate the stability or instability of time-variant system by its time-invariant model.

4. Quite often either the same periodically varying parameter or several synchronous parameter oscillations of different amplitudes appear in the mathematical system description. Here the synchronous parameters are meant to be periodic parameters having the same fixed oscillation frequencies and phases. In those cases, the transfer function numerator of the stationarized oscillatory object takes an operator form which order is a number of the similar items. For example, if the mathematical description of the linear periodical nonstationary system is given by the equation

$$a_3 \ddot{x} + [a_2 + a(t)]\ddot{x} + a_1 \dot{x} + [a_0 + a(t)]x = 0,$$

then the stationarized system description is

$$a_3 s^3 + a_2 s^2 + a_1 s + a_0 = -s^2 \frac{a}{2} e^{-j\varphi} - \frac{a}{2} e^{-j\varphi}$$

or

$$W^{-1}(s) = \frac{a_3 s^3 + a_2 s^2 + a_1 s + a_0}{s^2 + 1} = -\frac{a}{2} e^{-j\varphi}$$

or

$$W(s) = \frac{s^2 + 1}{a_3 s^3 + a_2 s^2 + a_1 s + a_0} = -\frac{2}{a} e^{-j\varphi}.$$

The constant part of the parameter variation profile is included in the stationary part of the system here.

To simulate the system behavior numerically the parameter by the highest operator degree term should be left stationary. It can be done by dividing all the terms of the governing equation by this parameter. Otherwise, the transfer function's numerator and denominator powers become equal and the simulation results are distorted owing to numerical discretization of the processes.

The following paragraph provides examples of parametric resonance condition evaluation for the systems described by various differential equations.

4.4 Examples

Here, we provide the example of the first parametric resonance excitation condition evaluation for different types of systems.

(a)

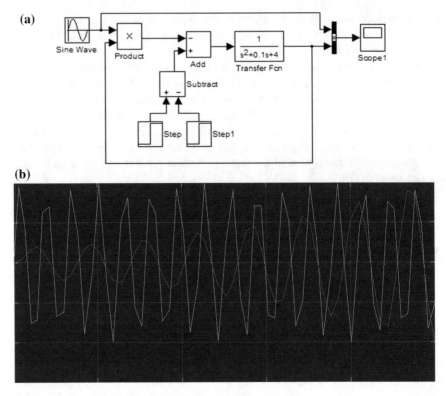

(b)

Fig. 4.6   a Numerical model diagram, **b** excitation process, $a = 0.5$

Example 1.   Let us consider a periodically nonstationary system described by the equation

$$\ddot{x} + 0.1\dot{x} + (4 + a\sin\Omega t)x = 0,$$

where $W^{-1}(s) = s^2 + 0.1s + 4$.

Based on modulus equality condition (4.20) at $s = j\Omega/2$ and $a_0 = 0$

$$\sqrt{\left(4 - \frac{\Omega^2}{4}\right)^2 + 0.01\frac{\Omega^2}{4}} = \frac{a}{2}.$$

We can derive now the lower excitation amplitude boundary evaluation for the first parametric resonance, that is $a_{min} = 0.4$ the eigenfrequency of the stationary part being $\omega_0 = 2$ rad/s and the parameter oscillation frequency $\Omega = 2\omega_0 = 4$ rad/s. Figure 4.6 shows the numerical model diagram and simulation process to the first parametric resonance excitation for Simulink.

Fig. 4.7 Excitation region boundaries for Example 2

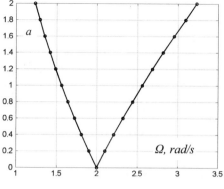

Fig. 4.8 Excitation region boundaries for Example 3

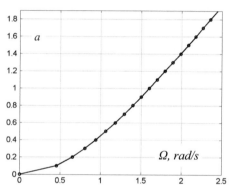

Example 2. Let us consider a periodically nonstationary system described by the equation

$$\ddot{x} + \dot{x}a \sin \Omega t + x = 0.$$

Here $W^{-1}(s) = (s^2 + 1)/s$. It follows from the condition (4.20) that $a = \left|4 - \Omega^2\right|/\Omega$.

Figure 4.7 shows the calculated excitation region boundary.

Example 3. Let a periodically nonstationary system be described by the equation

$$\ddot{x} + (\dot{x} + x)a \sin \Omega t = 0.$$

From which we derive $W^{-1}(s) = \frac{s^2}{s+1}$. Condition (4.20) gives $a = \frac{\Omega^2}{\sqrt{4+\Omega^2}}$ (see Fig. 4.8).

Chapter 5
Parametric Resonances of the Second and Higher Orders

5.1 The Second Parametric Resonance

As before, let a T-periodic variable parameter be described as

$$a(t) = a_0 + a \sin \Omega(t - \tau)$$

and its input signal takes the form

$$x(t) = x_0 + \tilde{x}(t) = x_0 + A \sin \Omega t, \tag{5.1}$$

here x_0 and $\tilde{x}(t)$ are the constant and variable components of the same frequency Ω, correspondingly. The output parametric element signal is derived as

$$y(t) = a(t)x(t) = a_0 x_0 + a_0 A \sin \Omega t + a x_0 \sin \Omega(t - \tau) + a A \sin \Omega t \sin \Omega(t - \tau)$$

$$= a_0 x_0 + a_0 A \sin \Omega t + a x_0 \sin \Omega(t - \tau) + \frac{a A}{2} \cos \Omega \tau + \frac{a A}{2} \cos(2\Omega t - \Omega \tau).$$

Separating the constant and variable components of the frequency Ω and ignoring the last term, we can see that

$$y(t) \approx y_0 + \tilde{y}(t) = a_0 x_0 + \frac{a A}{2} \cos \Omega \tau + a_0 \tilde{x}(t) + \frac{a x_0}{A} \tilde{x}(t - \tau). \tag{5.2}$$

This implies

$$y_0 = a_0 x_0 + \frac{a A}{2} \cos \Omega \tau. \tag{5.3}$$

$$\tilde{y}(t) = a_0 \tilde{x}(t) + \frac{a x_0}{A} \tilde{x}(t - \tau). \tag{5.4}$$

© Springer International Publishing AG, part of Springer Nature 2017
L. Chechurin and S. Chechurin, *Physical Fundamentals of Oscillations*,
https://doi.org/10.1007/978-3-319-75154-2_5

It follows directly from the structural system representation in Fig. 4.4 that the constant components x_0 and y_0 are connected with each other through the static gain $W(0)$ in the stationary part as

$$x_0 = -W(0)y_0, \quad W(0) = \frac{H(0)}{G(0)}. \tag{5.5}$$

Substituting (5.3) into (5.5) the constant component takes the form

$$x_0 = -\frac{aA\cos\Omega\tau}{2}\frac{W(0)}{1+a_0W(0)} = -\frac{aA\cos\Omega\tau}{2}W_{cl}(0). \tag{5.6}$$

Using (5.6), the variable component (5.4) can be obtained now as

$$\tilde{y}(t) = a_0\tilde{x}(t) - \frac{a^2\cos\Omega\tau}{2}W_{cl}(0)\tilde{x}(t-\tau). \tag{5.7}$$

Engaging the symbolic complex notation, we can derive now the following ratio:

$$\frac{\tilde{Y}(j\Omega)}{\tilde{X}(s)} = a_0 - \frac{a^2}{2}W_{cl}(0)e^{-j\Omega\tau}\cos\Omega\tau.$$

Taking into account $\Omega\tau = \varphi$ and $\cos\Omega\tau = (e^{j\Omega\tau}+e^{-j\Omega\tau})/2$, the transfer function of the periodically nonstationary parameter with respect to the second parametric resonance oscillations is expressed as

$$W(j2\varphi) = a_0 - \frac{a^2}{4}W_{cl}(0) - \frac{a^2}{4}W_{cl}(0)e^{-2\varphi}. \tag{5.8}$$

The frequency excitation conditions (4.18) and (4.19) to the second parametric resonance oscillations become

$$W^{-1}(j\Omega) = -W(j2\varphi) \tag{5.9}$$

or

$$W(j\Omega) = -W^{-1}(j2\varphi). \tag{5.10}$$

As in the case of the first parametric resonance, T-periodic parameter transfer function (5.8) does not depend on the amplitude A, but the phase φ of the coordinate oscillations relative to the parameter oscillations. It is a circle centered at the point $u_0 = a_0 - a^2W_{cl}(0)/4$ with a radius $\rho_0 = a^2W_{cl}(0)/4$. As in the case of the first parametric resonance, the two first terms of the transfer function (5.8) can be attributed to the stationary part of the subsystem, for example, by rewriting the excitation condition (5.9) as a modulus balance

$$\left|W_e^{-1}(j\Omega)\right| = a^2 W_{cl}(0)/4,$$

$$W_e(j\Omega) = \frac{H_{cl}(j\Omega)}{G_{ck}(j\Omega) + u_0 H_{cl}(j\Omega)}, \tag{5.11}$$

$$u_0 = a_0 - \frac{a^2}{4} W_{cl}(0).$$

As in first parametric resonance case, the second parametric resonance excitation conditions are determined by the frequency characteristics of the stationary part of the system. But the similarities end at this point.

Unlike the first parametric resonance, the second parametric resonance transfer function (5.8) remains legitimate with respect to the coordinate oscillations of the frequency, that is equal to the parameter variation frequency Ω. The important difference is that the second parametric resonance oscillations are fundamentally asymmetric because of the presence of the constant component (5.6). Moreover, the constant component of the oscillations depends on the amplitude A, and grows with the grows of the amplitude once the resonance is excited. It is interesting that the constant component of oscillations x_0 emerges even if the parameter variation profile does not have any constant term, in other words, $a_0 = 0$. Any attempts to eliminate the constant component x_0, result in the vanishing of the second parametric resonance. This happens if the stationary subsystem filtrates the constant component and $H(0) = 0$. For the same reason, the second parametric resonance does not appear if the periodic parameter is chained with the coordinate differentiating components. It is also interesting to know that the parameter varying at the oscillation frequency does not pass that frequency and just "a cooperation" with the constant component allows the oscillations of frequency Ω to make it through the parametric element.

Example 1. Let a periodically nonstationary system be described by the operator equation

$$(s^2 + 0.1s + 1)x + (s + 5)xa \sin \Omega t = 0.$$

In this case,

$$W^{-1}(p) = \frac{s^2 + 0.1s + 1}{s + 5},$$

$$W_e^{-1}(j\Omega) = W^{-1}(j\Omega) + \frac{5}{4}a^2,$$

$$H(0) = 5, \quad G(0) = 1, \quad a_0 = 0, \quad W_{cl}(0) = 5.$$

Condition (5.11) takes the form

$$\left|\frac{4}{5a^2} \frac{1 - \Omega^2 + j0.1\Omega}{5 + j\Omega} + 1\right| = 1.$$

Fig. 5.1 Example. The
second parametric resonance
excitation conditions

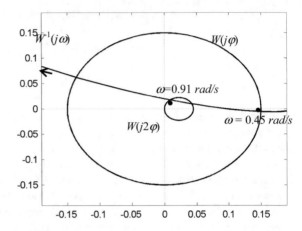

Fig. 5.2 Example. The
second parametric resonance
oscillogram

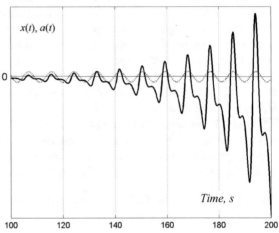

Figure 5.1 illustrates the second parametric resonance excitation conditions at the frequency $\Omega = 0.91$ rad/s and the amplitude $a = 0.3$.

It is evident from Fig. 5.1 that the second parametric resonance will take place since the point $\omega = \Omega = 0.91$ rad/s sits inside the circle of the second parametric resonance. Simultaneously, the point $\omega = \Omega/2 = 0.45$ rad/s does not sit inside the first parametric resonance circle, so there are no oscillations of $\Omega/2$ frequency. For the boundary point, the second parametric resonance excitation oscillogram has the form depicted in Fig. 5.2 As it can be seen, x-coordinate oscillations are asymmetric under the symmetric parameter variation plotted by the dotted line. It is worth paying attention to the fact that in contrast to the first parametric resonance, the coordinate amplitude growth is nonlinear.

Example 2. Consider the problem of unstable equilibrium stabilizing of an ideal inverted pendulum by the vertical vibration its joint (Kapitza's pendulum, see [10]).

The inverted pendulum equations for the case of vibrating its rotation axis differ from those of an ordinary sustained pendulum just by the sign at the acceleration g.

$$p^2 - \frac{g - a\Omega^2 \sin \Omega t}{l} = 0$$

Therefore, we write the frequency parametric resonance excitation condition (4.19) as follows:

$$\frac{1}{(j\Omega/2)^2 - g/l} = \frac{2l}{ja\Omega^2} e^{j\varphi}.$$

From which we derive the first parametric resonance excitation condition in the following form

$$a\Omega < \frac{l\Omega}{2} + \frac{2g}{\Omega}.$$

It is the first stabilization condition of the inverted pendulum.

The second condition also results from the same operator equation for the inverted pendulum. For the case of the second parametric resonance at $s = j\Omega = j\omega$, the frequency equation is

$$(j\Omega)^2 - g/l = W_2(j\varphi),$$

where $W_2(j\varphi)$ is the circle-shaped amplitude–phase characteristic of the same variable parameter $(a\Omega^2/l) \sin \Omega t$ with respect to the second parametric resonance oscillations. The rough approximation of the circle radius is

$$|W_2(j\varphi)| \approx 2(a\Omega^2/l)^2.$$

Substituting the approximation into the frequency equation from modulus equality, we define the reduced second stabilization condition

$$a\Omega > \sqrt{2gl}.$$

Combining both the conditions, we obtain the well-known stability conditions in respect of the unstable inversed pendulum equilibrium:

$$\sqrt{2gl} < a\Omega < \frac{l\Omega}{2} + \frac{2g}{\Omega}.$$

Table 5.1 lists the transfer functions for the first and second parametric resonances as applied to several simple periodic parameter variation profiles.

Table 5.1 First and second parametric resonance approximation for various profiles of parameter change

$a(t)$	$W_{1P}(j\varphi)$	$W_{2P}(j\varphi)$
$1 + a\sin\Omega t$	$1 - \frac{a}{2}e^{-j\varphi}$	$1 - \frac{a^2}{4}W_{CL}(0) + \frac{a^2}{4}W_{CL}(0)e^{-j2\varphi}$
$1 + a\cos\Omega t$	$1 - \frac{a}{2j}e^{-j\varphi}$	$1 - \frac{a^2}{4}W_{CL}(0) - \frac{a^2}{4}W_{CL}(0)e^{-j2\varphi}$
$1 + a\mathrm{sign}\cos\Omega t$	$1 - \frac{2a}{\pi}e^{-j\varphi}$	$1 - \frac{4a^2}{\pi^2}W_{CL}(0) + \frac{4a^2}{\pi^2}W_{CL}(0)e^{-j2\varphi}$
$1 + a\mathrm{sign}\sin\Omega t$	$1 - \frac{2a}{\pi j}e^{-j\varphi}$	$1 - \frac{4a^2}{\pi^2}W_{CL}(0) - \frac{4a^2}{\pi^2}W_{CL}(0)e^{-j2\varphi}$
$\begin{cases} 1, \ nT \leq t \leq (n+\gamma)T \\ 0, \ (n+\gamma)T \leq t \leq (n+1)T \\ \gamma \in (0,1) \end{cases}$	$\gamma - \frac{\sin\pi\gamma}{\pi}e^{-j\pi\gamma}e^{-j\varphi}$	$\gamma - \frac{\sin^2\pi\gamma}{\pi^2}W_{CL}(0)$ $+ \left(\frac{\sin^2\pi\gamma}{\pi^2}W_{CL}(0) - \frac{\sin 2\pi\gamma}{2\pi} \right)e^{-j2(\pi\gamma+\varphi)}$

5.2   Periodically Nonstationary Systems with Multifrequency Parameter

Let T-periodic parameter $a(t)$ follow an arbitrary variation profile. As it is well known, the parameter signal can be represented as a Fourier series in harmonic components as

$$a'_m \sin m\Omega t + a''_m \cos m\Omega t, \quad m = 0, 1, 2, \dots, . \tag{5.12}$$

The component can be rewritten in the dense form

$$a_m \sin(m\Omega t + \beta_m), \tag{5.13}$$

where the amplitude and phase are denoted as

$$a_m = \sqrt{a'^2_m + a''^2_m}, \quad \text{and} \quad \beta_m = \operatorname{arctg}\frac{a''_m}{a'_m}. \tag{5.14}$$

With the harmonic component (5.13), the following first parametric resonance transfer function can be written down using formula (4.13):

$$W(jm\varphi) = -\frac{|a_m|}{2j}e^{j\beta_m}e^{-jm\varphi} = -\frac{a'_m + ja''_m}{2j}e^{-jm\varphi} = -\rho_m e^{-jm\varphi},$$

$$r = |\rho_m| = \frac{a_m}{2} = \frac{\sqrt{a'^2_m + a''^2_m}}{2} \tag{5.15}$$

According to (5.18), the second parametric resonance transfer function at the m th harmonic of the parameter is derived in a similar way:

$$W(j2m\varphi) = -\frac{|a_m|^2}{4} W_{cl}(0) \left[1 + e^{-j2m\varphi} e^{j2\beta_m} \right] = -W_{cl}(0) \left(|\rho_m|^2 + \rho_m^2 e^{-j2(m\varphi - \beta_m)} \right). \quad (5.16)$$

The conditions to the first and second parametric resonances at the m th harmonic of the parameter in the plane of the inverse stationary part frequency characteristic, for example look like

$$W^{-1}\left(j\frac{m\Omega}{2} \right) = -W(jm\varphi) \qquad (5.17)$$

$$W^{-1}(jm\Omega) = -W(j2m\varphi). \qquad (5.18)$$

If the periodic parameter exists under more than one coordinate derivatives, in other words in terms like

$$s^{(i)}[a(t)x(t)], \quad i = 1, 2, \ldots, \qquad (5.19)$$

then with regard to the first parametric resonance either the transfer functions for each of the separate terms can be evaluated and summed up then, or the transfer function of all the terms can be evaluated at once. With the second parametric resonance, a number of singularities have to be taken into account. First, for the second parametric resonance, it is necessary to take into account the oscillations with a frequency $m\Omega$ of the first parametric resonance. It can appear at the double frequency of parameter variation, i.e., at the frequency $2m\Omega$. In that case, the transfer function (5.15) is modified and becomes

$$W(j2m\varphi) = -\frac{a_m'^2 + a_m''^2}{4} W_{c\ell}(0) - \left[\frac{(a_m' + ja_m'')^2}{4} W_{c\ell}(0) + \frac{a_m' + ja_m''}{2j} \right] e^{-j2m\varphi}. \quad (5.20)$$

We already urged to remember that for the terms like (5.19), the second parametric resonance does not appear. And finally, if there is the sum of the several synchronously varying parameters of the same frequency but a different phase, the transfer function of the parameters is not just the sum of the transfer functions of the terms as are as the second parametric resonance is considered. In other words, the superposition principle of linear systems is not true for the second parametric resonance oscillations in linear periodically nonstationary systems; and in addition, this is another distinctive feature of the second parametric resonance. In such a situation, all the synchronous items are reasonable to be reduced to the same argument prior to the transfer function calculation just as it was doing in transformation of (5.12)–(5.13).

It remains to stress that the Fourier series coefficients can be found as follows:

$$a_0 = \rho_0 = \frac{1}{T} \int_0^T a(t) dt \tag{5.21}$$

$$a'_m = \frac{2}{T} \int_0^T a(t) \sin m\Omega t \, dt \tag{5.22}$$

$$a''_m = \frac{2}{T} \int_0^T a(t) \cos m\Omega t \, dt, \tag{5.23}$$

or in the complex form

$$a_m = a'_m + ja''_m = \frac{2j}{T} \int_0^T a(t) e^{-jm\Omega t} dt. \tag{5.24}$$

One can also use a handbook. The coefficient ρ_m of the parameter transfer function $W_m(\varphi)$ and the appropriate parametric resonance circle radiuses $r = |\rho_m|$ take the form

$$\rho_m = \frac{a_m}{2j} = \frac{1}{T} \int_0^T a(t) e^{-jm\Omega t} dt \tag{5.25}$$

and agree with Fourier complex series expansion coefficients of the periodic parameter function.

Example. Let the periodically nonstationary system

$$10\ddot{x} + 0.01\dot{x} + 40x + ax \, \text{sgn} \sin \Omega t = 0.$$

be excited by the parameter

$$a(t) = a\text{sgnsin}(\Omega t).$$

In this case,

$$a_1 = \frac{4a}{\pi}, \quad a_3 = \frac{a}{3\pi}, \quad W^{-1}(s) = 10s^2 + 0.01s + 40.$$

Figure 5.3 illustrates the calculation of the first parametric resonance excitation conditions under the third harmonic at $a = 1$.

The circle radiuses of the first and second parametric resonances are 2/3 and 0.05, correspondingly. Figure 5.4 shows the third parametric resonance oscillogram where

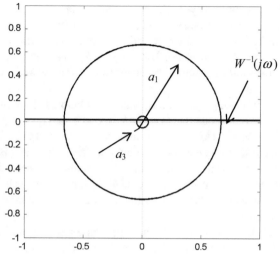

Fig. 5.3 Example. Evaluation of the parametric resonance excitation conditions at the third harmonic

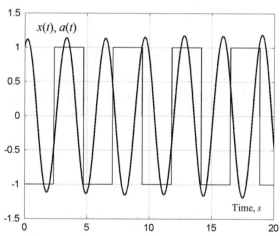

Fig. 5.4 Example. The third parametric resonance excitation (the first parametric resonance at the third harmonic)

the resonance is well observed based on the frequency relation between the parameter oscillation (in the form of square-sine function with a frequency of 4/3 rad/s) and the output coordinate.

5.3 Stationarization of Pulse Dynamic System

The essential feature of digital control systems is the time quantization/discretization of control processes. The quantization generally happens in the devices that connect the continuous control objects with control computers and also in continuous-to-discrete coordinate converters. The time quantization operation that is performed by time-pulse keys is the linear amplitude-pulse modulation of a continuous signal. In practice the pulse modulation is neglected at a high sampling frequencies and a digital system is then treated as a continuous one. To understand and assess what kind of phenomena are lost and gained as a result of that neglection we need to know what kind of features the quantization process adds to the dynamics of the digital system.

For the input amplitude-pulse modulation, the impulse key forms the output pulse of the amplitude that is equal to the instantaneous input signal at the make time of the key. The closing time is assumed to be equal to T. Thus, the key is a periodically nonstationary element and a pulse system with a key is a periodically nonstationary one. Idealization requires that the impulse key delivers the pulses of infinitesimal durations. A pulse area is equal to the input signal value at the moment of key make. It means the parameter variation profile is a T-periodic delta function of unit area denoted $\delta_T(t)$.

The T-periodic delta function Fourier expansion has a flat frequency spectrum, where all the harmonics have the same amplitudes, in accordance with the first relation of (5.25)

$$\rho_m = \frac{a_m}{2j} = \frac{1}{T} \int_0^T \delta_T(t) e^{-jm\Omega t} dt = \frac{1}{T}. \tag{5.26}$$

The first parametric resonance circles (4.13) coincide at the fundamental frequency $\Omega = 2\pi/T$ and the harmonic frequency, or

$$W(jm\varphi) = \frac{1}{T}(1 - e^{-jm\varphi}). \tag{5.27}$$

The second parametric resonance circles (5.26)

$$W(j2m\varphi) = \frac{1}{T}\left\{\left[1 - \frac{1}{T}W_{cl}(0)\right] + \left[\frac{1}{T}W_{cl}(0) - 1\right]e^{-j2m\varphi}\right\},$$

$$W_{cl}(0) = \frac{W(0)}{1 + T^{-1}W(0)}$$

are enveloped by the first parametric resonance circles. The function, that is the inverse of the key transfer function (5.27), is the vertical line through the point $u = -T/2$. Thus, the condition of the absence of the parametric excitation in the pulse system at any quantization frequency takes a simple visual form on the Nyquist hodograph plane (see Fig. 5.5):

$$\text{Re } W(j\omega) > -T/2, \quad 0 \le \omega < \infty. \tag{5.28}$$

The straight line $u = -T/2$ travels along the real axis from minus infinity to the origin, thus covering the entire left half-plane if the period varies from infinity to

Fig. 5.5 Parametric
excitation evaluation of the
idealized pulse system

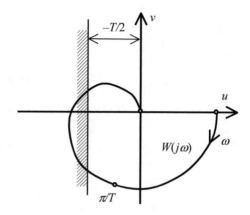

zero. Therefore, the rigid condition of the absence of the single-frequency parametric
resonance excitation is that the Nyquist hodograph for the stationary part lies in the
right half-plane.

From the interval range of the quantization frequencies $\Omega = 2\pi/T$, the point
excitation conditions should be checked at the fundamental and harmonic frequencies

$$\text{Re } W\left(\frac{jm\Omega}{2}\right) > -T/2, \quad m = 1, 2, \dots. \tag{5.29}$$

This condition is easy to visualize. The disadvantage is the variable right side of
inequality (5.29), which can be written as

$$\text{Im } [s\,W(s)]_{s=jm\Omega/2} > -m\pi/2. \tag{5.30}$$

The first parametric resonance excitation condition includes the excitation conditions
at the harmonics, if the following inequality is true in the low half-plane

$$|\text{Im } s\,W(s)|_{s=j\Omega/2} \geq m^{-1}\,|\text{Im } s\,W(s)|_{s=jm\Omega/2}. \tag{5.31}$$

This condition restricts a rise rate to the imaginary part of the modified frequency
characteristic $j\omega W(j\omega)$ in the low half-plane to limit, thereby the stationary part
resonance properties.

The effect of an idealized impulse element has been considered. The mathematical
model of systems with ideal impulse element is very important in the stability analysis
of nonlinear dynamic systems with discontinuous and piecewise characteristics. Now,
let us turn to the influence of a real impulse element.

A real impulse element has a finite pulse duration and the pulse amplitude is
proportional to the input value at the make moments of the key. A real key has
different descriptions depending on the fixation duration and the type of a pulse
former. Let us take the elementary real impulse element to fix an instantaneous input

Fig. 5.6 Parametric
excitation evaluation
concerning real pulse
systems

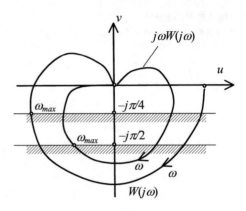

signal value over an entire quantization period. Such kind of a key is represented as
a series connection formed of an ideal impulse element and a clamper with a transfer
function

$$W_f(z, s) = \frac{z - 1}{z} \cdot \frac{1}{s}, \quad z = e^{sT}.$$

The gain of the clamper at $s = jm\Omega/2$ is

$$W_f(jm\Omega/2) = \frac{4}{jm\Omega}, \quad m = 1, 3, 5, \ldots$$

and taking into account the ideal key gain (5.27), the stability loss condition takes
the form

$$\text{Re } s^{-1}W(s)\big|_{s=jm\Omega/2} > -T/4.$$

This condition has the following equivalent forms

$$\text{Re } \frac{4}{jl\Omega} W(jm\Omega/2) > -\frac{T}{2}, \quad m = 1, 3, 5 \ldots \tag{5.32}$$

or

$$\text{Im } W(jm\Omega/2) > -m\pi/4 \tag{5.33}$$

which can be inspected by the frequency representations, given in Fig. 5.6.

In this case, the inequality that is similar to the condition (5.32) is

$$|\text{Im } W(s)|_{s=j\Omega/2} \geq m^{-1} |\text{Im } W(s)|_{s=jm\Omega/2}. \tag{5.34}$$

Let us consider several simple examples.

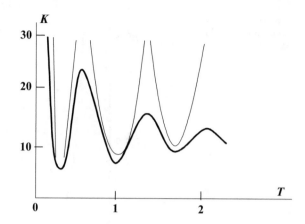

Fig. 5.7 Example 1. Excitation regions of parametric resonances of ideal pulse system

Example 1. Ideal pulse oscillator. Let the system include the oscillatory unit

$$W(s) = \frac{k}{(s + \alpha)^2 + \beta^2},$$

with the ideal impulse key in the feedback loop. The approximate excitation condition is derived from condition (5.28) as follows:

$$k \geq \frac{1}{2\pi T}(\pi m^2 - \alpha^2 T^2 - \beta^2 T^2) + \frac{2\pi T m^2 \alpha^2}{\pi m^2 - \alpha^2 T^2 - \beta^2 T^2}, \quad m > \frac{T}{\pi}\sqrt{\alpha^2 + \beta^2}.$$

It is interesting to compare this approximate condition to the exact solution obtained using the *z*-transformation

$$W(z) = Z\left[\frac{k}{\beta}\frac{\beta}{(p + \alpha)^2 + \beta^2}\right] = \frac{k}{\beta}\frac{zd \sin \beta T}{z^2 - 2zd \cos \beta T + d^2}.$$

The characteristic equation of the closed pulse system

$$z^2 - 2zd \cos \beta T + d^2 + \frac{k}{\beta}zd \sin \beta T = 0$$

results in the solution by $z = \pm 1$:

$$k = \pm\frac{\beta}{d \sin \beta T}\left(1 \pm 2d \cos \beta T + d^2\right), \quad d = e^{-\alpha T}.$$

Figure 5.7 compares the approximate solution (dotted line) and the exact one.

Example 2. Real pulse oscillator. Let us consider the same oscillator with the real impulse key represented by the ideal impulse element and clamper. The imaginary part of the continuous part frequency response characteristic is

Fig. 5.8 Example 2.
Parametric excitation
boundaries

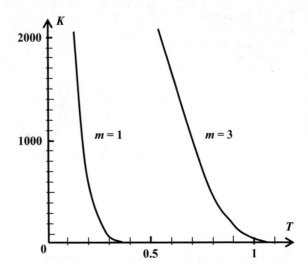

$$\text{Im } W(j\omega) = -\frac{2\alpha\omega k}{(\alpha^2 + \beta^2 - \omega^2)^2 + 4\alpha^2\omega^2}.$$

It follows from the condition (5.33) that

$$k = \frac{T}{8\alpha m}\left[\left(\alpha^2 + \beta^2 - \frac{\pi^2 m^2}{T^2}\right)^2 + 4\alpha^2 \frac{\pi^2 m^2}{T^2}\right].$$

Figure 5.8 illustrates the function $k(T)$ for the high frequencies $\omega^2 = (\pi m/T)^2 > \alpha^2 + \beta^2$ at $\alpha = 1$, $\beta^2 = 83$, $m = 1, 3$.

Chapter 6
Higher Order Parametric Systems

As it follows from Chap. 4, the single-frequency parametric oscillation excitation depends rather on the natural frequencies and the resonance pick values of the frequency response than on the order of the mathematical model of the system. Let us consider a system with two natural frequencies.

6.1 Systems with Two Natural Frequencies

Example. Let the operator equation of a simple two-frequency system have the form

$$(s^2 + 0.1s + 1)(s^2 + 0.1s + 4)x + (a \sin \Omega t)x = 0$$

and the inverse transfer function is

$$W^{-1}(s) = (s^2 + 0.1s + 1)(s^2 + 0.1s + 4).$$

The next equality is obtained from (4.20)

$$a^2 = \frac{1}{64} \left[(4 - \Omega^2)^2 + 0.04\Omega^2 \right] \left[(16 - \Omega^2)^2 + 0.04\Omega^2 \right].$$

With a fixed, there can be four positive solutions of the equation. This means that the first parametric resonance excitation region can consist of two subregions. It can be clearly seen from the relative location of hodographs at one of the amplitudes a, shown in Fig. 6.1. Figure 6.2 displays the region boundary obtained in the graphical form for a number of amplitudes.

© Springer International Publishing AG, part of Springer Nature 2017
L. Chechurin and S. Chechurin, *Physical Fundamentals of Oscillations*,
https://doi.org/10.1007/978-3-319-75154-2_6

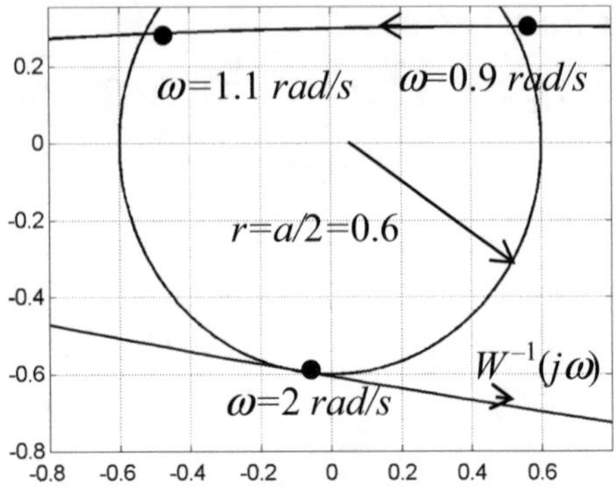

Fig. 6.1 Example. Graphic
determination of the
boundaries

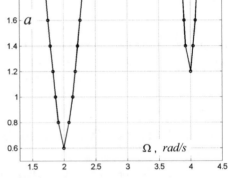

Fig. 6.2 Example.
Excitatory region boundary

6.2 Distributed Parameter Systems

As follows from Chap. 2, the frequency characteristics of ideal distributed systems
can have an infinite set of natural frequencies. In real systems, resonance picks
diminish as frequencies grow. It legitimates the limitation of stability analysis by
a certain number of natural frequencies. When digitalizing, it limits the number of
units chosen to model the distributed system. Let us perform the stability evaluation
and simulation of a pendulum system oscillations as an example. The pendulum
suspension is modeled as the unit of distributed stiffness that is absolutely rigid for
bending. In that case, the parameter $l(t)$ includes distributed parameters such as mass,
m_i, and stiffness, c_i, a coordinate control loop consists of a pendulum model with
lumped inertia moment, J_0, and length, l_0. The problem of the parametric oscillations
of such a pendulum is similar to some degree to the old problems of the oscillation
of a spring carriage coupling or a mine cage.

Let us represent the approximate model of a distributed parameter as a train of n
units. Each unit has the following description:

$$m_i \ddot{x} + \delta \dot{x} = f(n) - f(n+1)$$
$$f(n+1) = c_i[x(n) - x(n+1)].$$

(6.1)

Let us assume the boundary conditions at the suspension point as $x = 0$: $f(0)$, $x(0) = 0$. Having applied z-transform, we arrive at

$$c_i(z - 1)X(z) + zF(z) = zf(0)$$
$$(m_i p^2 + \delta p)X(z) - (1 - z)F(z) = zf(0)$$
(6.2)

or

$$F(z, p) = \frac{zf(0)[z - (-1 + 2ch\xi)]}{c_i(z^2 - 2zch\xi + 1)}$$

$$X(z, p) = -\frac{zf(0)}{c_i(z^2 - 2zch\xi + 1)}$$

where we defined $ch\xi = \frac{m_i s^2 + \delta s}{2c_i} + 1$.

Returning back from z-transform to the function origin, the oscillation equation for the n th unit, $n = 1, 2, \dots N$, takes the form

$$X(s, N) = -\frac{sh\,\xi N}{c_i(sh\,\xi)ch\,\xi N}F(s, N), \quad n = 1, 2, \dots N.$$

From this, the transfer function $W(s, N)$ and phase-amplitude characteristic $W(j\omega, N)$ to the end unit become as

$$W(s, N) = \frac{X(s, N)}{F(s, N)} = -\frac{sh\xi N}{c_i(sh\xi)ch\xi N}.$$
(6.3)

Let us consider the train of 4 units as an example, $N = 4$. After a number of trigonometry manipulations, we can derive the following expression:

$$W(s, 4) = -\frac{(ch\xi)(ch\xi - \sqrt{0.5})(ch\xi + \sqrt{0.5})}{c_i(ch\xi - \sqrt{0.75})(ch\xi - \sqrt{0.25})(ch\xi + \sqrt{0.75})(ch\xi + \sqrt{0.25})}.$$
(6.4)

Let us evaluate the result numerically. Let the physical parameters of the system are the same as in the previous lumped example, i.e., $m_0 = 1$ kg, $l_0 = 1$ m. Let the distributed parameter values are taken as follows: $c_i = 6$ kg/m, $m_i = 0.25$ kg, $\delta = 0.01$ kg/s. The amplitude–phase–frequency characteristic $W(j\omega, 4)$ has four resonance frequencies. At the second resonance frequency ($\omega = 5$ rad/s), the modulus of characteristic is equal to 12 and the lag $\tau = 0.5$ s.

Figure 6.3 displays the 4-unit spring pendulum model and the parametric resonance excitation simulation results. According to the transfer function (6.4), the parametric loop includes four dynamic units of the second order.

$$Fcn2 = \frac{2s^2 + 0.1s + 100}{2s^2 + 0.1s + 13} \qquad Fcn3 = \frac{2s^2 + 0.1s + 30}{2s^2 + 0.1s + 50}$$

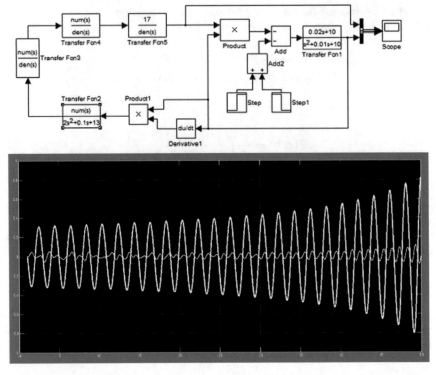

Fig. 6.3 The Simulink model and the simulation results for the multiunit spring pendulum (initial unit pulse duration is 0.1 s)

$$Fcn4 = \frac{2s^2 + 0.1s + 170}{2s^2 + 0.1s + 190} \quad Fcn5 = \frac{17}{2s^2 + 0.1s + 150}$$

The parameter oscillation amplitude is proportional to the square of coordinate oscillation amplitude. Therefore, the system becomes unstable at large deviations of the coordinate. The results illustrate the main physical idea: the parametric resonance emerges in the system because the distributed parameters add some lag.

Part III
Parametric Resonance in Nonlinear System Oscillations

Chapter 7 is an introduction and reviews briefly the method for one-frequency-based analysis of self-oscillations, and forced and vibrational oscillations of non-linear systems. Well-known describing function method or one-frequency harmonic linearization method is the main instrument. The conditions of parametric resonance excitation, which were derived in the previous part, enable the one-frequency analysis of stability loss in nonlinear oscillations in Chap. 8. Finally, Chap. 9 presents the analysis of entrainment and synchronization of nonlinear oscillation and approximate conditions of these phenomena.

Chapter 7
Nonlinear System Oscillations: Harmonic Linearization Method

The method of describing functions is mainly used to analyze nonlinear dynamic systems in a single-frequency harmonic approximation. The method is also known as the method of harmonic linearization or harmonic balance method or the first harmonic method. The method directly arises from spectral and asymptotic methods as their first approximation. For the first time, the method was applied in 1934 by the Soviet scientist V. Kotelnikov to evaluate the performance of self-excited oscillation generators.

7.1 Harmonic Linearization Factors

Let the input harmonic

$$x(t) = x_0 + \tilde{x} = x_0 + A \sin \psi, \quad \psi = \omega t \tag{7.1}$$

be fed to the nonlinear element $F(x)$. The output periodic signal from a nonlinear element has the form

$$y(t) = F(A \sin \omega t) \tag{7.2}$$

and can be expanded into Fourier series

$$F(A \sin \omega t) = F_0 + \sum_{m=1}^{\infty} (f'_m \sin m\psi + f''_m \cos m\psi), \tag{7.3}$$

where according to formulations (5.21) to (5.24), the coefficients are known to be

© Springer International Publishing AG, part of Springer Nature 2017
L. Chechurin and S. Chechurin, *Physical Fundamentals of Oscillations*,
https://doi.org/10.1007/978-3-319-75154-2_7

$$F_0 = \frac{1}{2\pi} \int_0^{2\pi} F(x_0 + A \sin \psi) d\psi, \tag{7.4}$$

$$f'_m = \frac{1}{\pi} \int_0^{2\pi} F(x_0 + A \sin \psi) \sin m\psi \, d\psi, \tag{7.5}$$

$$f''_m = \frac{1}{\pi} \int_0^{2\pi} F(x_0 + A \sin \psi) \cos m\psi \, d\psi. \tag{7.6}$$

Let us distinguish the basic harmonic component of the output periodic signal of the nonlinear element. This is the first harmonic in most cases

$$y_1(t) = F_0 + f'_1 \sin \psi + f''_1 \cos \psi. \tag{7.7}$$

The ratios of the components F_0, f'_1, f''_1 of the output signal $y_1(t)$ to the constant component x_0 and amplitude A of the input signal are written as

$$q_0(A, x_0) = \frac{F_0}{x_0} = \frac{1}{2\pi x_0} \int_0^{2\pi} F(x_0 + A \sin \psi) d\psi, \tag{7.8}$$

$$q'_1(A, x_0) = \frac{f'_1}{A} = \frac{1}{\pi A} \int_0^{2\pi} F(x_0 + A \sin \psi) \sin \psi \, d\psi, \tag{7.9}$$

$$q''_1(A, x_0) = \frac{f''_1}{A} = \frac{1}{\pi A} \int_0^{2\pi} F(x_0 + A \sin \psi) \cos \psi \, d\psi. \tag{7.10}$$

The expressions (7.8)–(7.10) are called harmonic linearization factors with respect to the constant, in-phase and quadrature components, correspondingly. They can be given the complex form too

$$W_1(A, x_0) = q'_1(A, x_0) + j q''_1(A, x_0), \tag{7.11}$$

where

$$W_1(A, x_0) = \frac{1}{\pi A} \int_0^{2\pi} F(x_0 + A \sin \psi) e^{-j\psi} d\psi. \tag{7.12}$$

The complex harmonic linearization factor is also called as nonlinear element transfer function.

In the general case, the idea of the harmonic linearization method is the replacement of a nonlinear element by a linear one; the complex transfer constant of the latter depends on the amplitude A and constant component, x_0, of the input harmonic, i.e.,

$$F(x) \cong W_1(A, x_0)x. \tag{7.13}$$

The presence of the imaginary part of the harmonic linearization factor $q_0' \neq 0$ means that the nonlinear element adds a phase shift to the output harmonic signal of the same frequency to transform the input harmonic of frequency ω to the output signal. In contrast to the periodically nonstationary element, the phase shift and modulus of the harmonic transfer gain of the nonlinear element depend on the input amplitude.

The harmonic linearization factor description (7.11) can be found in the literature in the operator form

$$W_1(A, x_0) = \mathrm{Re}W(A, x_0) + \frac{s}{\omega}\mathrm{Im}W(A, x_0). \tag{7.14}$$

We should be careful here since the harmonic linearization factors are derived in the assumption of the harmonic input and output signals. It means that the expression (7.14) is legitimate at $s = j\omega$ only.

It is worth mentioning that the nonlinear element static (with respect to the constant component) gain q_0 is zero as long as the input signal does not have a constant component ($x_0 = 0$). It is also true in the case of odd-symmetrical nonlinearity characteristic, i.e., $F(x) = -F(-x)$. The imaginary harmonic linearization factor q_1', is zero if the nonlinearity $F(x)$ is single valued. Moreover, if the nonlinearity is symmetric and does not depend on derivatives, the definition of the harmonic linearization factor becomes simpler as follows:

$$W_1(A) = \mathrm{Re}W_1(A) = \frac{4j}{\pi A} \int_0^{\pi/2} F(A \sin \psi)e^{-j\psi} d\psi. \tag{7.15}$$

7.2 Examples of Harmonic Linearization Factor Evaluation

Cubic function. The symmetric cubic function

$$F(x) = cx^3 \tag{7.16}$$

is well known from Duffing equation. The real harmonic linearization factor of the function is determined by formulation (7.15) as follows:

$$W_1(A) = q_1'(A) = \frac{4}{\pi A} \int_0^{\pi/2} cA^3 \sin^3 \psi \sin \psi \, d\psi = \frac{4cA^2}{\pi} \int_0^{\pi/2} \sin^4 \psi \, d\psi$$

$$= \frac{cA^2}{\pi} \int_0^{\pi/2} \left(1 - 2\cos 2\psi + \frac{1 + \cos 4\psi}{2} \right) d\psi = \frac{3cA^2}{4}. \tag{7.17}$$

Ideal relay characteristic. The nonlinear characteristic of an ideal three-position relay has the definition

$$F(x) = \begin{cases} B \operatorname{sgn} x, & |x| > b \\ 0, & |x| < b. \end{cases} \tag{7.18}$$

Figure 7.1 shows the function $F(A \sin \psi)$ plotted for the input signal $x(t) = A \sin \psi$. Due to the symmetry of the single-valued characteristic (7.18), the harmonic linearization factor is derived using the formulation (7.15) in the following way:

$$W_1(A) = q_1'(A) = \frac{4}{\pi A} \int_0^{\pi/2} F(A \sin \psi) \sin \psi \, d\psi$$

$$= \frac{4}{\pi A} \int_{\psi_1}^{\pi/2} B \sin \psi \, d\psi = \frac{4B}{\pi A} \cos \psi_1. \tag{7.19}$$

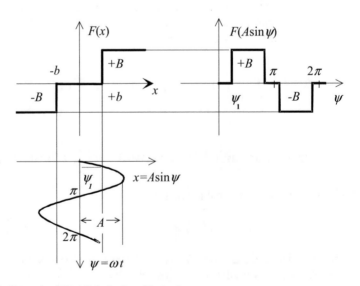

Fig. 7.1 Example of $W_1(A)$ derivation. Ideal relay

The equality $A \sin \psi_1 = b$ arises directly from Fig. 7.1 and then

$$W_1(A) = \frac{4B}{\pi A}\sqrt{1 - \frac{b^2}{A^2}}. \tag{7.20}$$

The expression for the harmonic linearization factor becomes simpler in the case of a perfect two-position relay ($b = 0$):

$$W_1(A) = \frac{4B}{\pi A}. \tag{7.21}$$

Saturation-type nonlinearity. The nonlinearity is described by the expression

$$F(x) = \begin{cases} x, & |x| < b \\ B \operatorname{sgn} x, & |x| > b. \end{cases} \tag{7.22}$$

Figure 7.2 presents the characteristic $F(x)$ and the output signal $F(A \sin \psi)$ of the nonlinear element. The harmonic linearization factor is derived as follows:

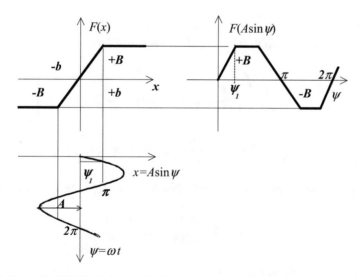

Fig. 7.2 Example of $W_1(A)$ derivation. Real relay

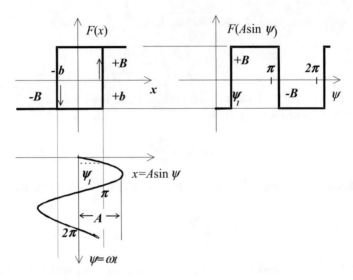

Fig. 7.3 Example of $W_1(A)$ derivation. Two-position relay

$$W_1(A) = q_1'(A) = \frac{4}{\pi A} \int_0^{\pi/2} F(A \sin \psi) \sin \psi \, d\psi$$

$$= \frac{4}{\pi A} \left(\int_0^{\psi_1} \frac{B}{b} A \sin \psi \sin \psi \, d\psi + \int_{\psi_1}^{\pi/2} B \sin \psi \, d\psi \right) \qquad (7.23)$$

$$= \frac{2B}{\pi b} \left(\psi_1 + \frac{\sin 2\psi_1}{2} \right), \quad \psi_1 = \arcsin \frac{b}{A}.$$

Real two-position relay characteristic.   Figure 7.3 displays the nonlinear function

$$F(x) = \begin{cases} -B \operatorname{sgn} \dot{x}, & |x| < b \\ +B \operatorname{sgn} x, & |x| > b \end{cases} \qquad (7.24)$$

and the periodic function $F(A \sin \psi)$.

Since the double-valued characteristic is symmetric, the coefficients (7.9) and (7.10) are derived in the phase variation range from zero to $\psi = \pi$ as follows:

$$q_1(A) = \frac{2}{\pi A} \int\limits_0^\pi F(A \sin \psi) \sin \psi \, d\psi$$

$$= \frac{2}{\pi A} \left[-\int\limits_0^{\psi_1} B \sin \psi \, d\psi + \int\limits_{\psi_1}^\pi B \sin \psi \, d\psi \right] \tag{7.25}$$

$$= \frac{4B}{\pi A} \cos \psi_1 = \frac{4B}{\pi A} \sqrt{1 - \frac{b^2}{A^2}} = \frac{2B}{\pi b} \sin 2\psi_1,$$

$$q_1'(A) = \frac{2}{\pi A} \int\limits_0^\pi F(A \sin \psi) \cos \psi \, d\psi$$

$$= \frac{2B}{\pi A} \left[-\int\limits_0^{\psi_1} \cos \psi \, d\psi + \int\limits_{\psi_1}^\pi \cos \psi \, d\psi \right]$$

$$= -\frac{4B}{\pi A} \sin \psi_1 = -\frac{4Bb}{\pi A^2} = -\frac{4B}{\pi b} \sin^2 \psi_1,$$

where $\psi_1 = \arcsin \frac{b}{A}$, or in the complex form

$$W_1(A) = \frac{4B}{\pi A} \left(\sqrt{1 - \frac{b^2}{A^2}} - j \frac{b}{A} \right). \tag{7.26}$$

The harmonic linearization factors for most frequent nonlinear elements are listed in Table 7.1.

7.3 Self-excited Oscillations

Let us consider the autonomous dynamic system given in Fig. 7.4. The system has the description in the operator form

$$G(s)x(t) + H(s)F[x(t)] = 0, \tag{7.27}$$

Fig. 7.4 Autonomous system with nonlinearity

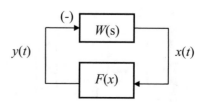

Table 7.1 Harmonic linearization factors for popular nonlinearities

Nonlinearity	Harmonic linearization factors	Fourier coefficients
	$q_1' = \dfrac{4K_p}{\pi A}\sqrt{1-\left(\dfrac{b}{A}\right)^2}\ (A \geq b)$; $q_1'' = 0$	$a_0 = \dfrac{4K_p}{\pi A}\left(\dfrac{b}{A}\right)^2 \Big/ \sqrt{1-\left(\dfrac{b}{A}\right)^2}$; $\|\rho_1\| = \dfrac{2K_p}{\pi A}\dfrac{\left\|2(b/A)^2 - 1\right\|}{\sqrt{1-(b/A)^2}}$
	$q_1' = \dfrac{4K_p}{\pi A}\sqrt{1-\left(\dfrac{b}{A}\right)^2}\ (A \geq b)$; $q_1'' = -\dfrac{4K_p b}{\pi A^2}$	$a_0 = \dfrac{2K_p}{\pi A} \Big/ \sqrt{1-\left(\dfrac{b}{A}\right)^2}$; $\|\rho_1\| = \dfrac{2K_p}{\pi A} \Big/ \sqrt{1-\left(\dfrac{b}{A}\right)^2}$
	$q_1' = K$ $q_1'' = \dfrac{4K_p}{\pi A^2}$	$a_0 = K + j\dfrac{2K_p}{\pi A}$; $\|\rho_1\| = \dfrac{2K_p}{\pi A}$
	$q_1' = \dfrac{K}{\pi}\left[\dfrac{\pi}{2} + \arcsin\left(1-\dfrac{2b}{A}\right)\right.$ $\left.\ + 2\left(1-\dfrac{2b}{A}\right)\sqrt{\dfrac{b}{A}\left(1-\dfrac{b}{A}\right)}\right]$; $q_1'' = -\dfrac{4K_p b}{\pi A}\left(1-\dfrac{b}{A}\right)$; $A \geq b$	$a_0 = \dfrac{K}{\pi}\left[\dfrac{\pi}{2} + \arcsin\left(1-\dfrac{2b}{A}\right)\right.$ $\left.\ + 2\sqrt{\dfrac{b}{A}\left(1-\dfrac{b}{A}\right)} - j\dfrac{2b}{A}\right]$; $\|\rho_1\| = \dfrac{2K}{\pi}\dfrac{b}{A}$
	$q_1'' = \dfrac{2K}{\pi}\left[\arcsin\dfrac{b_1}{A} - \arcsin\dfrac{b_0}{A} +\right.$ $\left.\ + \dfrac{b_1}{A}\sqrt{1-\left(\dfrac{b_1}{A}\right)^2} - \dfrac{b_0}{A}\sqrt{1-\left(\dfrac{b_0}{A}\right)^2}\right]$; $q_1'' = 0$; $A \geq b$	$a_0 = \dfrac{2K}{\pi}\left(\arcsin\dfrac{b_1}{A} - \arcsin\dfrac{b_0}{A}\right)$; $\|\rho_1\| = \dfrac{2K}{\pi}\left[\dfrac{b_1}{A}\sqrt{1-\dfrac{b_1}{A}} - \dfrac{b_0}{A}\sqrt{1-\dfrac{b_0}{A}}\right]$

Fig. 7.5 Low-pass filter

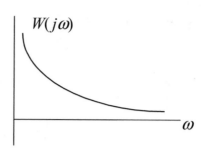

Fig. 7.6 Resonance filter

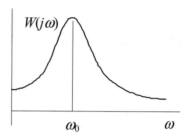

where $W(s) = H(s)/G(s)$ is the traditional notation of the transfer function of the linear stationary part of the system.

The use of the harmonic linearization method is based on the well-known filter hypothesis. It assumes that any real linear dynamic subsystem is the low-pass filter, i.e.,

$$|W(j\omega)| \gg |W(jn\omega)|, \quad n = 2, 3, 4, \ldots \tag{7.28}$$

and the resonance filter at all frequencies, i.e.,

$$|W(j\omega_0)| \gg |W(j\omega)|, \tag{7.29}$$

where ω_0 is the resonance frequency.

The properties are illustrated by the frequency characteristics given in Figs. 7.5 and 7.6.

In that case, the higher harmonic components are supposed to be filtered out in the linear part of the system. It legitimates the procedure of harmonic approximation.

Let us find self-oscillation excitation conditions. The harmonically linearized equation is written in the form

$$G(j\omega) + H(j\omega)W_1(A) = 0. \tag{7.30}$$

If we approach the equation analytically, we split it into imaginary and real equations. Their composition defines two unknown quantities: the amplitude A and frequency ω of the self-excited oscillations.

Fig. 7.7 Nyquist hodograph

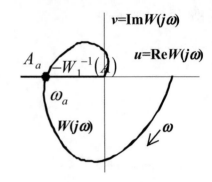

Fig. 7.8 Inverse Nyquist hodograph

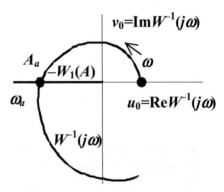

If there is a constant component g_0 in the system input or the nonlinearity is asymmetric ($F_0 \neq 0$), Eq. (7.30) is studied together with the complementary static equation obtained from (7.30) at $s = j\omega = 0$:

$$G(0)x_0 + H(0)q_0(A, x_0)x_0 = H(0)g_0. \tag{7.31}$$

The graphical derivation of the unknown A and ω employs the Nyquist hodograph

$$W(j\omega) = -W_1^{-1}(A), \quad \text{where } W(j\omega) = H(j\omega)/G(j\omega). \tag{7.32}$$

The cross-point (A_s, ω_s) of Nyquist hodograph and the inverse complex harmonic linearization factor define the problem solution as shown in Fig. 7.7. The similar solution is attained in the inverse Nyquist hodograph plane

$$W^{-1}(j\omega) = -W_1(A), \tag{7.33}$$

as illustrated in Fig. 7.8.

Example. Let us evaluate the self-excited oscillation parameters for a relay system. Let the self-oscillatory system have the following transfer function of its linear part

$$W(s) = \frac{k}{s(T_1 s + 1)(T_2 s + 1)}$$

and the nonlinear element is a relay

$$F(x) = B \, \text{sgn} \, x.$$

Using harmonic linearization factor (7.21) and frequency equality (7.33), the relation can be given the following form:

$$-j T_1 T_2 \omega^3 - (T_1 + T_2)\omega^2 + j\omega = -\frac{4Bk}{\pi A}.$$

From this, we can derive two equalities in respect to the real and imaginary parts of the equation as follows:

$$(T_1 + T_2)\omega^2 = \frac{4Bk}{\pi A}$$
$$\omega(1 - T_1 T_2 \omega^2) = 0.$$

The final result is

$$\omega^2 = \frac{1}{T_1 T_2}$$
$$A = \frac{4Bk}{\pi} \left(\frac{T_1 T_2}{T_1 + T_2} \right).$$

7.4 Forced Oscillations

The harmonic linearization method is successfully employed for forced oscillation analysis. Let the departure point be the description of a system in the operator form

$$G(s)x(t) + H(s)F[x(t)] = R(s)v(t), \tag{7.34}$$

where $v(t)$ is an input signal. Let us introduce new input signal as follows:

$$x_{in}(t) = \frac{R(s)}{H(s)} v(t). \tag{7.35}$$

The initial equation is rewritten in the form then

$$G(s)x(t) + H(s)F[x(t)] = H(s)x_{in}(t). \tag{7.36}$$

Having used the transfer function notation $W(s) = H(s)/G(s)$ of the open-loop linear part, Eq. (7.36) is given the form

$$1 + W(s)F[x(t)] = W(s)x_{in}(t) \tag{7.37}$$

and the block diagram for such a system can be shown in a traditional manner (Fig. 7.9).

Harmonically linearized equation (7.37)

$$1 + W(s)W_1(A)x(t) = W(s)x_{in}(t) \tag{7.38}$$

can be written in the form

$$\frac{x(t)}{x_{in}(t)} = \frac{W(s)}{1 + W(s)W_1(A)}. \tag{7.39}$$

Let us give the input and output signals the complex form

$$x_{in}(t) = A_{in}e^{-j\omega t} \tag{7.40}$$

$$x(t) = Ae^{-j(\omega t + \phi)}. \tag{7.41}$$

Then, Eq. (7.39) with $s = j\omega$ implies

$$\frac{A_{in}}{A}e^{-j\phi} = W^{-1}(j\omega) + W_1(A). \tag{7.42}$$

Figure 7.10 visualizes the expression (7.42).

For a given amplitude A and frequency $\omega = \omega'$ of the forced oscillations, the required amplitude of the input A_{in} can be defined by the vector $A\omega'$. The vector origin is at the point A belonging to the complex (in general case) harmonic linearization factor $-W_1(A)$. The vector modulus is A_{in}/A and the vector phase ϕ is the frequency characteristic argument of the closed-loop system. As A_{in} grows from zero to infinity, the absence of the intersection of the circle $(A_{in}/A)\exp(-j\phi)$ with the inverse Nyquist hodograph is replaced by, at first, their tangency point and then the two intersection points ω' and ω''.

As mentioned above, the balance between the phases and amplitudes in the closed-loop system is the oscillation excitation boundary condition concerning the linear stationary and periodically nonstationary systems. The phase balance takes place

Fig. 7.9 Block diagram for forced oscillation system with nonlinearity

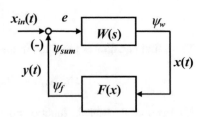

Fig. 7.10 Harmonic balance
for the forced oscillations

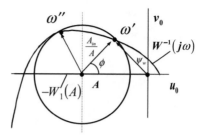

also in the nonlinear systems when the total phase shift in the closed-loop system is zero:

$$\psi_w + \psi_f + \psi_{sum} = 0, \tag{7.43}$$

where the phase shifts of the linear part ψ_w, the nonlinear element ψ_f, and the summing member ψ_{sum} are denoted as in Fig. 7.10. The amplitude balance means that the total gain of the open-loop system is equal to 1, i.e.,

$$|W(j\omega)W_1(A)W_{sum}(A_{in})| = 1. \tag{7.44}$$

The summing member transfer function $W_{sum}(A_{in})$ is introduced in (7.44) in addition to the already known Nyquist hodograph $W(j\omega)$ and the complex harmonic linearization factor $W_1(A)$. The summing member transfer function depends on the input signal $A_{in}\exp(j\omega t - j\phi)$ and the complex harmonic linearization factor (see Fig. 7.10) as follows:

$$W_{sum}(A_{in}) = \frac{e}{AW_1(A)} = \frac{A_{in}e^{-j\phi} - AW_1(A)}{AW_1(A)}.$$

The balance between phases and amplitudes is a general oscillation excitation condition. Thus, for an autonomous nonlinear system ($A_{in} = 0$), the balance conditions define the self-oscillation excitation region boundaries. According to Nyquist stability criterion, the balance conditions define the stability boundary for the linear stationary autonomous system ($W_1(A) = 1, A_{in} = 0$). At last, there are no oscillations in all the cases, if the system has positive stability margin in respect to phase and amplitude both. Let us consider several examples.

Example 1. Let the elementary nonlinear system consist of the linear integrator with nonlinear quadratic negative feedback

$$W(s) = k/s, \ F(x) = x^2 \operatorname{sgn} x.$$

According to (7.9), the feedback harmonic linearization factor is

Fig. 7.11 Example 1.
Forced oscillation evaluation

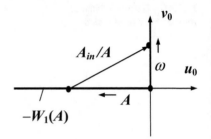

$$W_1(A) = \frac{8A}{3\pi}.$$

Equation (7.42) has a simple form

$$\frac{A_{in}}{A}e^{-j\phi} = \frac{8A}{3\pi} + j\frac{\omega}{k}$$

with a geometric interpretation given in Fig. 7.11.

It follows from the modulus equality

$$A^2 \approx \omega^2 k^{-2}(\sqrt{3A_{in}^2 k^4 \omega^{-4} + 1} - 1)$$

and the phase is derived in the form

$$\phi = -\arcsin\frac{\omega A}{k A_{in}}.$$

As mentioned above, the use of the harmonic linearization method is based on filter and resonance hypothesizes. However, there are numerous instances when those conditions are met but the harmonic linearization-based simulation displays its low accuracy comparing to physical experiments, even leads to qualitatively wrong conclusions. And on the contrary, there are examples in which the harmonic linearization gives good approximation of excitation conditions while neither of the above-mentioned hypothesis hold.

Example 2. Let us consider a nonlinear system

$$\ddot{x} + \dot{x} + 30F(x) = A_{in}\sin\omega t$$

$$F(x) = \begin{cases} x, & |x| < 1 \\ \text{sgn}x, & |x| > 1. \end{cases}$$

Its linear part is described by the transfer function

$$W(s) = \frac{30}{s(s + 1)}$$

that is essentially a low-pass filter.

According to (7.22), the harmonic linearization factor $F(x)$ is

$$W_1(A) = \left(\gamma + \frac{\sin \pi \gamma}{\pi}\right); \quad \gamma = \frac{2}{\pi} \arcsin \frac{1}{A}.$$

Let us derive the linearized frequency equation (7.42)

$$\frac{A_{in}}{A} = \left| W^{-1}(j\omega) + W_1(A) \right|.$$

Let us choose $\omega = 3$ rad/s and $A = 2.3$. We immediately obtain $A_{in} = 0.6$. At the same time, the actual solution of the nonlinear system gives $A = 5.0$. Moreover, the motion with an amplitude $A = 2.2$ corresponding to the solution $A_{in} = 0.6$ turns out to be unstable.

7.5 Vibration Linearization

There is a family of well-known applications of oscillation studies in engineering based on vibration linearization phenomenon. Shakers, vibrational screens, shaking containers, vibratory compactors have become routine means of field engineering long ago. The application of high-frequency vibrations to eliminate the harmful influence of nonlinearities on the dynamic systems process quality has also become known since long time ago. Relays, dry friction units, dead zones, backlashes, saturation units, etc., are among such kinds of nonlinearities. The effect of the nonlinearity influence elimination with the help of adding high-frequency oscillation is called as vibration linearization because the nonlinear element characteristic is smoothed and as if were linearized under the action of vibrations. Certainly, the vibration linearization is fundamentally different from numerous mathematical linearization tools that use expansions in Taylor, Fourier, Laurent, etc., series since the linearization has the nature of a physical effect.

Let the input signal with constant and variant components be

$$x = x_0 + \tilde{x} = x_0 + A \sin \omega t \tag{7.45}$$

and it is fed to the nonlinear element $F(x)$.

The approximation is made for small x_0 as

$$y = y_0 + \tilde{y} = F(x_0 + A \sin \omega t) \approx F(A \sin \omega t) + \left.\frac{dF(x)}{dx}\right|_{x=\tilde{x}} x_0. \tag{7.46}$$

Fig. 7.12 Vibration
linearization for two-position
relay

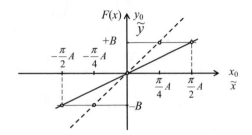

Following the main idea of the chapter, we are looking for the direct relations between the constant and variable components that are based on the harmonic linearization and emerge out of the approximation (7.46) in the forms

$$\tilde{y} = W_1(A)\tilde{x} \tag{7.47}$$

$$y_0 = a_0 x_0 = \left[W_1(A) + \frac{A}{2} \frac{dW_1(A)}{dA} \right] x_0. \tag{7.48}$$

Therefore, the nonlinear element is represented by the linear gain that depends on the vibration amplitude. It is true for the variable component, the small quantity component, and slow motion components of the input signal.

Using the harmonic linearization factor (7.21) to an ideal two-position relay as a nonlinearity, we can obtain from (7.47) and (7.48)

$$\tilde{y} = \frac{4B}{\pi A}\tilde{x} \quad \text{and} \quad y_0 = \frac{2B}{\pi A}x_0. \tag{7.49}$$

Figure 7.12 gives the characteristics $F(x)$, $\tilde{y}(\tilde{x})$, and $y_0(x_0)$.

Let us now release the restriction of the constant component smallness but keep the condition $x_0 < A$. The constant signal component $y_0(x_0)$ is derived now in accordance with Fig. 7.13, as

$$y_0 = \frac{1}{2\pi} \int_0^{2\pi} F(x_0 + A \sin \omega t)d\psi = \frac{B}{2\pi} \left(\int_0^{\pi+\psi_1} d\psi - \int_{\pi+\psi_1}^{2\pi-\psi_1} d\psi + \int_{2\pi-\psi_1}^{2\pi} d\psi \right)$$

$$= \frac{2B}{\pi}\psi_1 = \frac{2B}{\pi} \arcsin \frac{x_0}{A}, \quad y_0 < B, \ x_0 < A. \tag{7.50}$$

Fig. 7.13 Example.
Constant component
computation

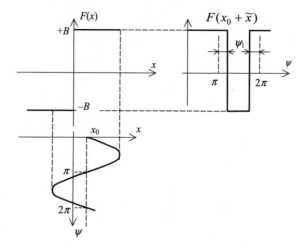

Fig. 7.14 Example.
Vibrational linearization

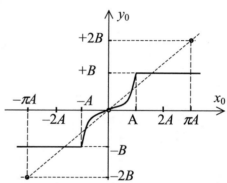

Figure 7.14 depicts the vibrationally linearized characteristic of the nonlinear element. The initial curve piece follows the straight line (7.49). In practice, the vibration frequency ω is chosen as high as to be filtered by the linear subsystem and thus it cannot be found at the system output. An external generator, natural self-oscillations, or a high-frequency system parameter can serve as vibration sources.

Chapter 8
Nonlinear System Oscillation Stability

One of the obvious effects of motion stability loss is known as jump resonance which occurs under certain situations in nonlinear dynamic systems.

The jump resonance condition is observed when the amplitude relation $A(A_{in})$ has the negative derivative dA/dA_{in} as it is evident from Fig. 8.1.

The critical points of the relation $A(A_{in})$ have the infinite derivative dA/dA_{in} or zero derivative dA_{in}/dA.

The jump resonance conditions arise from Eq. (7.42). Let us evaluate the modulus of Eq. (7.42) using the notation $W^{-1}(j\omega) = u_0 + jv_0$:

$$A_{in} = A\sqrt{[u_0 + W_1(A)]^2 + v_0^2}.$$

The derivative dA_{in}/dA has the form

$$\frac{dA_{in}}{dA} = \frac{[u_0 + W_1(A)]^2 + v_0^2 + A[u_0 + W_1(A)]dW_1(A)/dA}{\sqrt{[u_0 + W_1(A)]^2 + v_0^2}}.$$

The equality $dA_{in}/dA = 0$ defines the circle

$$[u_0 + u_{10}(A)]^2 + v_0^2 = r_{10}^2(A). \tag{8.1}$$

The circle radius and center are defined as

$$r_{10}(A) = \frac{A}{2}\left|\frac{dW_1(A)}{dA}\right|$$

$$u_{10}(A) = W_1(A) + \frac{A}{2}\left|\frac{dW_1(A)}{dA}\right|. \tag{8.2}$$

If the point of the inverse Nyquist hodograph is located inside the circle, the jump resonance happens. The following paragraph explains the jump resonance phenomenon as the stability loss due to the first parametric resonance excitation.

© Springer International Publishing AG, part of Springer Nature 2017
L. Chechurin and S. Chechurin, *Physical Fundamentals of Oscillations*,
https://doi.org/10.1007/978-3-319-75154-2_8

Fig. 8.1 Jump resonance

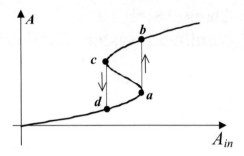

8.1 General Problem of Motion Stability

Let us consider a nonlinear nonautonomous system. Its operator notation is as follows:

$$G(s)x(t) + H(s)F[x(t)] = R(s)s(t). \tag{8.3}$$

Let us assume that a solution $x^*(t)$ exists. We will call it an unperturbed motion. Let us introduce a perturbed motion

$$x(t) = x^*(t) + \Delta x(t) \tag{8.4}$$

that emerges under the impact of a small external perturbation. Substituting (8.4) in Eq. (8.3) and subtracting the identity

$$G(s)x^*(t) + H(s)F[x^*(t)] = R(s)s(t), \tag{8.5}$$

we arrive at the autonomous system equation or incremental equation that describes the perturbed motion as

$$G(s)\Delta x(t) + H(s)\left\{F[x^*(t) + \Delta x(t)] - F[x^*(t)]\right\} = 0 \tag{8.6}$$

or

$$G(s)\Delta x(t) + H(s)\frac{\Delta F[x^*(t) + \Delta x(t)]}{\Delta x(t)}\Delta x(t) = 0 \tag{8.7}$$

and with $\Delta x(t) \rightarrow 0$

$$G(s)\Delta x(t) + H(s)\left[\frac{dF(x)}{dx}\right]_{x=x^*(t)}\Delta x(t) = 0. \tag{8.8}$$

This is the description of a linear nonstationary system because the derivative $\left[\frac{dF(x)}{dx}\right]_{x^*(t)}$ is a time-dependent function. Hence, if an incremental equation has a unique stable equilibrium, the unperturbed motion of the initial equation is also stable. This means that *the problem of the linear nonstationary system equilibrium leads to the problem of the nonlinear system oscillation stability. For the first time, this statement was formulated by Russian scientist A. Lyapunov in the form of a theorem.*

Periodic motion stability. Let us assume the T-periodic motion $\tilde{x}(t)$ has been derived by solving Eq. (8.3). This motion stability is governed by the incremental equation of type (8.8):

$$G(s)\Delta x(t) + H(s)\left[\frac{dF(x)}{dx}\right]_{x=\tilde{x}(t)} \Delta x(t) = 0, \tag{8.9}$$

that is to say, a nonlinear nonstationary equation. Here, the derivative dF/dx of the periodic motion $\tilde{x}(t)$

$$a(t) = \frac{dF^*}{dx} = \left[\frac{dF}{dx}\right]_{x=\tilde{x}(t)} \tag{8.10}$$

is a periodically variable parameter. We restrict the diversity of nonlinearities $F(x)$ by the odd-symmetric characteristics hereafter. In this case, the period of derivative variation (8.10) is the half-period of the coordinate changing $\tilde{x}(t)$, i.e., $\Omega = 2\omega$. The Fourier series expansion coefficients of the parameter are derived based on dependence (5.24) in the form

$$a_m = a'_m + ja''_m = \frac{2j\omega}{\pi} \int_0^{\pi/\omega} \frac{dF^*}{dx} e^{-j2m\omega t} dt$$

$$= \frac{2j}{\pi} \int_0^{\pi} \frac{dF^*}{dx} e^{-j2m\psi} d\psi, \quad \psi = \omega t, \quad m = 1, 2, \tag{8.11}$$

Due to (5.15), the first parametric resonance excitation condition (5.17) can be given the stationarized form as

$$1 + W(jm\omega)[a_0 - \rho_m e^{-jm\varphi}] = 0, \quad \varphi = 2\omega\tau. \tag{8.12}$$

This means that the $2T$-periodic motion $\tilde{x}(t)$ becomes unstable as soon as at least one of the points $m\omega$, $m = 1, 2, \ldots$ of the inverse Nyquist hodograph is located inside the m th circle or first parametric resonance circle with a center at $-a_0$ and radius $r_m = |\rho_m| = |a_m/2j|$ at the m th harmonic of parameter variation (8.10). For simplicity, the linear part of the system is assumed to be stable, and there are no self-excited oscillations in the autonomous nonlinear system.

Example 1. Let us examine the elementary nonlinear element

$$F(x) = ax^3.$$

Under the influence of the periodic motion

$$\tilde{x}(t) = A \sin \omega t,$$

the periodic parameter takes the form

$$a(t) = \left[\frac{dF}{dx}\right]_{x=\tilde{x}(t)} = 3aA^2 \sin^2 \omega t.$$

Using expression (8.11)

$$a_0 = \frac{2j}{\pi} \int_0^\pi 3aA^2 \sin^2 \psi d\psi = \frac{3}{2} aA^2$$

$$a_1 = \frac{2j}{\pi} \int_0^\pi 3aA^2 \sin^2 \psi e^{-j2\psi} d\psi = \frac{3}{2j} aA^2.$$

Then, it is not difficult to deduce the stability loss conditions by the known transfer function $W(s)$.

The calculation of coefficients (8.11) is not a difficult exercise if the nonlinear function derivative is available. At the same time, a lot of real nonlinearities have singularities where the derivative discontinuities. Moreover, at the first step of determining the periodic motion $\tilde{x}(t)$, the harmonic linearization factors are evaluated, and a harmonic linearization is applied. For that reason, it seems reasonable to find the coefficients a_m by the harmonic linearization factors $W_m(A)$.

Relation between stationarization and linearization factors. The jump resonance conditions were obtained based on the above physically understandable reasoning. The statement that the jump resonance conditions are those for oscillation stability loss needs to be validated. Let us provide the validation with regard to the single-frequency solution

$$\tilde{x}(t) = A \sin \psi, \quad \psi = \omega t. \tag{8.13}$$

Here, the periodic parameter (8.10) has the form

$$a(t) = \frac{dF(A \sin \psi)}{dA \sin \psi}. \tag{8.14}$$

The first parameter harmonic with a frequency $\Omega = 2\omega$ is assumed to be as a fundamental one. The average value of the parameter and its expansion coefficient according to the expression (5.25) are obtained as follows:

$$a_0 = \frac{1}{T} \int_0^T a(t)dt = \frac{1}{\pi} \int_0^\pi \frac{dF(A \sin \psi)}{dA \sin \psi} d\psi \qquad (8.15)$$

$$\rho_1 = \frac{1}{\pi} \int_0^\pi \frac{dF(A \sin \psi)}{dA \sin \psi} e^{-j2\psi} d\psi. \qquad (8.16)$$

To provide the connection of the parameter expansion coefficients with harmonic linearization factor (7.15), the derivative is evaluated as follows:

$$\frac{dW_1(A)}{dA} = -\frac{2j}{\pi A^2} \int_0^\pi F(A \sin \psi) e^{-j\psi} d\psi + \frac{2j}{\pi A^2} \int_0^\pi \frac{dF(A \sin \psi)}{dx} \frac{dx}{dA} e^{-j\psi} d\psi$$

$$= -\frac{W_1(A)}{A} + \frac{2j}{\pi} \int_0^\pi \frac{dF(A \sin \psi)}{dA \sin \psi} \sin \psi e^{-j\psi} d\psi$$

$$= -\frac{W_1(A)}{A} + \frac{1}{\pi} \int_0^\pi \frac{dF(A \sin \psi)}{dA \sin \psi} d\psi - \frac{1}{\pi} \int_0^\pi \frac{dF(A \sin \psi)}{dA \sin \psi} e^{-j2\psi} d\psi$$

$$= -\frac{W_1(A)}{A} + \frac{a_0}{A} - \frac{\rho_1}{A}$$

or

$$W_1(A) + A \frac{dW_1(A)}{dA} = a_0 - \rho_1. \qquad (8.17)$$

To derive the second connection equation, Eq. (7.15) is integrated by parts as

$$W_1(A) = -\frac{1}{\pi A} \int_0^{2\pi} F(A \sin \psi) d(e^{-j\psi})$$

$$= -\frac{1}{\pi A} \left[F(A \sin \psi) e^{-j\psi} \big|_0^{2\pi} - \int_0^{2\pi} \frac{F(A \sin \psi)}{d\psi} e^{-j\psi} d\psi \right]$$

$$= \frac{1}{\pi} \int_0^{2\pi} \frac{F(A \sin \psi)}{dA \sin \psi} \cos \psi e^{-j\psi} d\psi = \frac{1}{\pi} \int_0^\pi \frac{F(A \sin \psi)}{dA \sin \psi} (1 - e^{-j2\psi}) d\psi$$

or

$$W_1(A) = a_0 + \rho_1. \qquad (8.18)$$

Summing up and subtracting (8.17) and (8.18), the relationship equations take their final form

$$a_0 = W_1(A) + \frac{A}{2}\frac{dW_1(A)}{dA}$$
$$\rho_1 = -\frac{A}{2}\frac{dW_1(A)}{dA}. \tag{8.19}$$

The system (8.19) links the harmonic linearization factor and the harmonic stationarization coefficient to each other with respect to the first harmonic. The first equality in (8.19) defines the center of the parametric excitation circle; the circle radius is the modulus of the right side of the second equality. Thus, we can formulate the oscillation stability loss condition within the first harmonic approximation for the nonlinear system: the nonlinear system oscillations of a frequency ω become unstable if the inverse Nyquist hodograph point ω gets into the parametric resonance circle with a radius and a center taken from system (8.19). The derived equations (8.19) coincides with the jump resonance conditions (7.46). This means that the jump resonance is the form of stability loss of forced oscillations due to the first parametric resonance excitation.

Example 1. Ideal two-position relay Let us take a harmonic linearization factor in respect to the first harmonic (7.21). Its derivative has the form

$$\frac{dW_1(A)}{dA} = -\frac{2B}{\pi A^2}$$

and the following relations are derived from Eq. (8.19):

$$a_0 = \frac{2B}{\pi A}, \quad r_1 = |\rho_1| = \left|\frac{a_1}{2j}\right| = \frac{2B}{\pi A}. \tag{8.20}$$

Obviously, the radius and center of the excitation circle are the same and the circle crosses the origin. For the two-position relay dynamic system, the stability motion loss boundary due to the first parametric resonance excitation is determined by the condition

$$W^{-1}(j\omega) = -\frac{2B}{\pi A}(1 - e^{-j\varphi}) \tag{8.21}$$

in the plane of the inverse Nyquist hodograph. In the plane of the Nyquist hodograph, the circle on the right side of (8.21) becomes a vertical straight line crossing the right half-plane at the distance of $\pi A/4B$, and the oscillation stability condition looks like

$$\mathrm{Re}W(j\omega) \geq -\frac{\pi A}{4B}. \tag{8.22}$$

Figure 8.2a, b illustrates both the conditions.

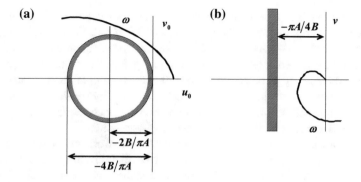

Fig. 8.2 Example 1. Oscillation excitation conditions in relay system

Example 2. Real two-position relay The derivative of the harmonic linearization factor derived from (7.25) and (7.26) is

$$\frac{dW_1(A)}{dA} = -\frac{4B}{\pi A^2 \cos\psi_1}(\cos 2\psi_1 - j\sin 2\psi_1)$$
$$= -\frac{4B}{\pi A^2 \cos\psi_1}e^{-j2\psi_1}$$

and Fourier coefficients (8.19) are

$$a_0 = \frac{2B}{\pi\sqrt{A^2 - b^2}}, \quad \rho_1 = \frac{2B}{\pi A\sqrt{A^2 - b^2}}e^{-j2\psi_1}. \tag{8.23}$$

Thus, motion stability loss condition (8.19) of a dynamic system with a real two-position relay characteristic can be given the form

$$\mathrm{Re}W(j\omega) \geq -\frac{\pi\sqrt{A^2 + b^2}}{4B} \tag{8.24}$$

and it is similar to the characteristics in Fig. 8.2.

8.2   Self-excited Oscillation Stability

Substituting the excitation circle equation (8.19) in first parametric resonance excitation condition (8.12), we can derive the condition for a self-excited oscillation stability loss as follows:

$$W^{-1}(j\omega) = -\left[W_1(A) + \frac{A}{2}\frac{dW_1(A)}{dA}\right] + \left[-\frac{A}{2}\frac{dW_1(A)}{dA}\right]e^{-j\varphi}. \tag{8.25}$$

The condition (8.25) is the same as the self-oscillation condition (7.33) at $\varphi = \pi$. This means that the inverse Nyquist hodograph $W^{-1}(j\omega)$, the negative harmonic linearization factor $-W_1(A)$, and the minus-signed first parametric resonance circle intersect at the same point (A_s, ω_s), as shown in Fig. 8.3. This is the consequence of the known theorem by A. A. Andronov of self-excited oscillations. This also means that we are not able to judge the self-oscillation stability by the first parametric resonance circle solely since the first parametric resonance excitation itself can be considered as a reason for the appearance of self-oscillations.

In most cases, the known simple reasoning allows to work out the self-oscillation stability. Let us assume that the self-excited oscillation amplitude A has been increased due to small external influence. We assume it has been moved from the point A_s to the point A_2 along the inverse harmonic linearization factor (see Fig. 8.3). If according to Nyquist criterion the point A_2 passes beyond the inverse Nyquist hodograph, the self-excited oscillation amplitude continues to grow, and the self-excited oscillations of an amplitude A_a become unstable. And on the contrary, as soon as the point A_2 enters the inverse Nyquist hodograph, the amplitude of A_2 will decrease and the self-excited oscillations will be stable.

It should be noticed that the location of the center of parametric resonance circles points the direction of an amplitude rise. Indeed, in accordance to the first equality of (8.13) the center $-a_0$ locates to the right from the point $-W_1(A_s)$, when the derivative $dW_1(A)/dA$ is negative, and the center lies to the left if the derivative is positive. Thus, the known and above-stated reasoning in respect of self-oscillation stability requires that the center of a parametric resonance circle belongs to the stability region to provide the self-excited oscillation stability. The center of parametric resonance circle is the mean value of the periodically variable parameter $a(t)$. Therefore, the above reasoning is only a necessary condition to the self-oscillation stability. It means the following: the self-oscillations of nonlinear system are stable if the linear stationary system with the averaged periodic parameter is also stable.

Fig. 8.3 Self-excited oscillation stability evaluation

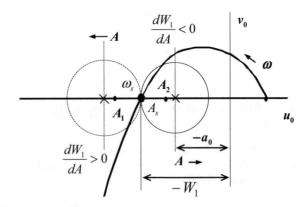

Fig. 8.4 Nyquist hodograph
of relay system

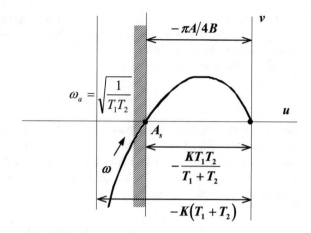

The additional information to this necessary condition on self-oscillation stability
can be derived from the second parametric resonance excitation conditions analysis
or the conditions of parametric resonances at the second or higher harmonics.

Example. Let us consider a specific example of self-oscillatory relay system from
Sect. 7.3. Nyquist hodograph of the problem takes the form

$$W(j\omega) = \frac{k}{j\omega(T_1 j\omega + 1)(T_2 j\omega + 1)}$$

and depicted in Fig. 8.4.

With an ideal two-position relay profile, the parametric resonance circles degen-
erate into the vertical straight lines $u = -\pi A /4B$ passing the self-oscillation point
(A_s, ω_s). Stability loss conditions (8.25) are not satisfied at the high frequencies
$\omega = m\omega_s, m = 1, 2, \ldots$ because the Nyquist hodograph modulus

$$|W(j\omega)| = \frac{k}{\omega\sqrt{(T_1^2\omega^2 + 1)(T_2^2\omega^2 + 1)}}$$

is a decay function. So in the context of single-frequency harmonic approximations,
the self-excited relay system oscillations are stable.

8.3 Forced Oscillation Stability

Let us consider simultaneously the forced oscillation existence condition (7.42) and
the parametric resonance excitation condition (8.19)

$$W^{-1}(j\omega) = -\left[W_1(A) + \frac{A_{in}}{A}e^{-j\phi}\right] \tag{8.26}$$

$$W^{-1}(j\omega) = -[W_1(A) + \frac{A}{2}\frac{dW_1(A)}{dA}] + \frac{A}{2}\frac{dW_1(A)}{dA}e^{-j\varphi}. \tag{8.27}$$

If the equation solution (8.26) takes place, that is to say, for example, the points ω' and ω'' in Fig. 8.5 (where $\dot{W}_1$ denotes the derivative with respect to A) are in the circle described by the right side of Eq. (8.27), the forced oscillations are unstable.

Since the sum of Eq. (8.19) is

$$a_0 + \rho_1 = W_1(A), \qquad (8.28)$$

the parametric resonance circle passes through the center of the circle described by the right side of Eq. (8.26), i.e., through the point $-W_1(A)$. That means, in turn, that Eqs. (8.26) and (8.27) do not have a common solution if the circle diameter of the parametric resonance excitation condition is less than the circle radius of the forced oscillation appearance condition:

$$A\left|\frac{dW_1(A)}{dA}\right| < \frac{A_{in}}{A} \quad \text{or} \quad A_{in} > A^2\left|\frac{dW_1(A)}{dA}\right|. \qquad (8.29)$$

That is the condition of the "absolute" stability of forced oscillation, i.e., their stability condition which does not depend on the form of the frequency characteristic and transfer function of the linear part of the system.

As well as for self-excited oscillations, the stability loss conditions of forced oscillations due to the first parametric resonance excitation can be corrected with the excitation conditions of the second parametric resonance and upper harmonic resonances (see Chap. 5).

Let us consider again two examples, in which the forced oscillation existence conditions were specified in Sect. 7.4 of Chap. 7.

Example 1. The problem of forced oscillation stability to the system with the integrator $W(s) = k/s$ and the quadratic feedback $F(x) = x^2\text{sgn}x$ is examined. As the harmonic linearization factor is $W_1(A) = 8A/3\pi$, it follows from (8.19) that

$$a_0 = \frac{12A}{3\pi}, \quad \rho_1 = \frac{a_1}{2j} = -\frac{4A}{3\pi}.$$

The parametric excitation circle

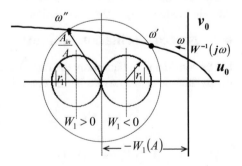

Fig. 8.5 Forced oscillation stability evaluation

$$(u_0 + a_0)^2 + v_0^2 = \left(\frac{4A}{3\pi}\right)^2,$$

where $u_0 + jv_0 = W^{-1}(j\omega) = 0 + j\omega k^{-1}$ cannot contain any point of the hodograph $W^{-1}(j\omega)$ in view of the following inequality:

$$\left(\frac{12A}{3\pi}\right)^2 + \frac{\omega^2}{k^2} > \left(\frac{4A}{3\pi}\right)^2.$$

The forced oscillations in the system are unstable. The result is obvious since the inverse Nyquist hodograph does not enter the left half-plane where there is the parametric resonance circle (see Fig. 7.11).

Example 2. A saturation type nonlinearity element is used in this example. Its harmonic linearization factor is

$$W_1(A) = \gamma + \frac{\sin \pi \gamma}{\pi}, \quad \gamma = \frac{2}{\pi} \arcsin \frac{1}{A},$$

and a_0 and ρ_1 follow from (8.19) as

$$a_0 = \gamma, \quad \rho_1 = \frac{\sin \pi \gamma}{\pi}.$$

For the forced oscillation amplitude chosen in the example, $A = 2.2$, they are $a_0 = 0.3$, $\frac{a_1}{2} = 0.26$, and the point $\omega = 3$ rad/s of the inverse Nyquist hodograph

$$W^{-1}(j\omega) = -0.033\omega^2 + j0.033\omega = -0.3 + j0.1$$

sits inside the parametric resonance circle

$$a_0 + \frac{a_1}{2} \exp(j\varphi) = -0.3 + 0.26 \exp(j\varphi),$$

as it is shown in Fig. 8.6.

8.4 Physical Interpretation of Oscillation Stability Loss

Let a phase–amplitude balance be kept in the nonlinear system shown in Fig. 7.9. The result is that the oscillations with a frequency ω and an amplitude A are assumed to appear and be stable in the system. In that case, the settled output of the nonlinear component is a periodic function of the periodic argument. The nonlinear element can be seen as a time-dependent dynamic gain, or the periodically varying gain of its input signal. Geometrically, the gain is the tangential line to the nonlinear characteristic, and therefore, this is a periodic parameter $a(t) = dF[x(t)]/dx(t)$, evaluated over the periodic motion $x = x^*$. It is known from Chap. 2 that the periodic parameter can introduce the phase shift to the dynamic system and also change the gain. If there

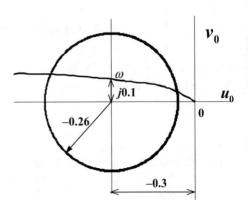

Fig. 8.6 Example 2.
Stability loss condition

is a specific phase shift φ between the periodic parameter and the input oscillations at which the phase shift and the gain change inserted by the periodic parameter are sufficient to reach a new phase–amplitude balance with some new periodic motion, those new oscillations emerge. The frequency and phase of those new (or secondary excited) oscillations are generally different from those of the fundamental (or primary settled) oscillations. In a special but frequent and principal case, the excited oscillations are the same in frequency with the fundamental oscillations and different in phase. That is the case when the first parametric resonance appears at the first/fundamental harmonic of a parameter variation. With a symmetric nonlinear characteristic, the parameter varies at the frequency $\Omega = 2\omega$, and the first parametric resonance leads to the parametric oscillation excitation at the frequency $\omega = \Omega/2$. The sum of the fundamental and new parametric oscillations of the same frequency forms the further oscillations different in phase and amplitude under the same frequency, i.e., the jump of the phase and amplitude of the fundamental oscillations is observed. The cause of those jumps or excited parametric oscillations is "hidden" because of this frequency coincidence. For that reason, various designations of the oscillation stability loss such as jump resonance, bending resonance, lake and island resonances, etc., have appeared in the literature. They all are based on one phenomenon that is parametric resonance.

A ferroresonance effect is well known in electrical engineering, and it is the best illustration of what has been said above. The effect is called ferroresonance because it happens in an *RLC* circle comprising an iron-core inductor (see Fig. 8.7a). The core inductor is known to have the nonlinear dependence of its inductance on an electric current. That is why the amplitude and phase jumps of AC current appear in the circle in Fig. 8.7a and the iron-core transformer circle in Fig. 8.7b under certain conditions. It is significant that the electromotive force which is transferred from the primary winding to the secondary circuit is equal to zero when there are two identical opposite connected secondary windings, as shown in Fig. 8.7c. And in spite of that, the input frequency oscillations are observed in the secondary circuit of Fig. 8.7c at the frequency of oscillation jumps of the circuit in Fig. 8.7b.

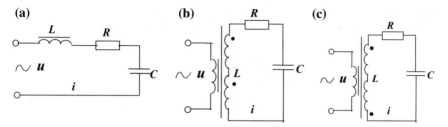

Fig. 8.7 Ferroresonance circuits

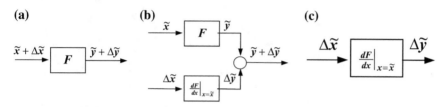

Fig. 8.8 Equivalent nonlinearity representation

These are the first parametric resonance oscillations separated from the fundamental oscillations.

Thus, having the steady fundamental primary oscillations, the nonlinear system becomes linear but periodically nonstationary concerning the secondary parametric oscillations. This means the nonlinear element $F(x)$ can be divided into two parts as it presented in Fig. 8.8: (1) for the fundamental primary oscillations $\tilde{x}$ and (2) for the secondary parametric oscillations $\Delta \tilde{x}$.

The second part has been singled out from the nonlinear element (it can be called secondary nonlinearity). It is presented in Fig. 8.8c. There are all the reasons to take the secondary nonlinearity as real, because it can be not only evaluated (see Fig. 8.9) but also measured by the device designed according the block diagram, given in Fig. 8.10.

In all the cases, the secondary nonlinearities or the so-called quasilinear characteristics depend on the secondary parametric oscillation phase but not on the amplitude. To verify the foregoing, let us calculate the harmonic linearization factor to the symmetric quasilinear characteristic $F(x)$, depicted in Fig. 8.11

$$\frac{j}{\pi A} \int_0^{2\pi} |A \sin(\psi + \varphi)| e^{-j(\psi + \varphi)} d\psi$$

$$= \frac{4j}{\pi} \int_0^{\pi/2} \frac{e^{l(\psi + \varphi)} - e^{-j(\psi + \varphi)}}{2j} e^{-j(\psi + \varphi)} d\psi$$

$$= 1 + \frac{2j}{\pi} e^{-j\varphi} = W_1(j\varphi).$$

Fig. 8.9 Introduced
nonlinearity evaluation

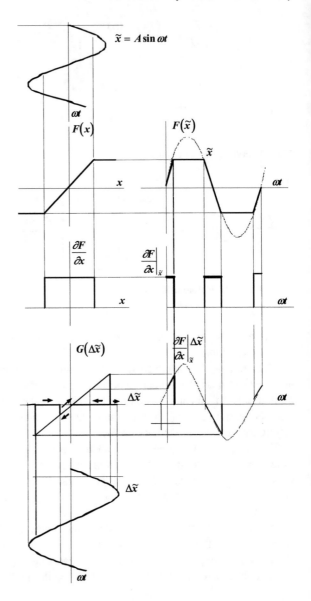

One can see that the harmonic linearization factor does not depend on the amplitude A and coincides with the first parametric resonance circle to the periodic parameter $a(t) = 1 + \mathrm{sgn}\,\sin\,\Omega t$ (see Table 5.1).

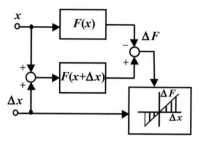

Fig. 8.10 The measuring of introduced nonlinearity

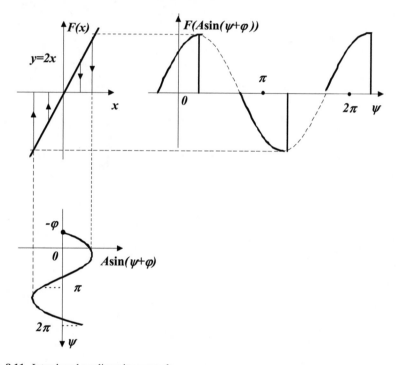

Fig. 8.11 Introduced nonlinearity example

Chapter 9
Synchronization of Oscillations

The synchronization of oscillatory systems is widely used in technology. The desynchronizing of an energy-producing electric machine with its grid or several AC machines with each other become a reason for emergency shutoffs of the plants and blackouts. The stable image in a television raster is impossible without a synchronizing phasing system of horizontal and vertical sweeps, etc.

Let us agree on several terms and definitions at least in the scope of the book prior to pass on the examination of synchronization effects. Let two harmonic signals of an equal frequency be called synchronous or frequency-synchronized. Moreover, if the relative phase of the signals is constant, such signals are called in-phase or phase-synchronized. Strictly speaking, if the relative phase of the signals is variable, the signal frequencies do not coincide. However, if the relative phase or so-called "hunting" phase varies over a period of time so that to be constant on average, the frequencies can be coincident, i.e., similar signals satisfy the concept of synchronous but not in-phase ones. To reduce synchronous oscillations to in-phase ones, special phase-synchronized loop control systems are used in technology. To provide the acceptable operation of synchronous electric machines, the synchronism between the machine rotation and network oscillations is mandatory. The terminology is very important here because the chapter deals with the analysis of synchronization effects at multiple frequencies when the signals with not equal but multiple frequencies become phase-synchronized, that is to say, nonsynchronous in-phase.

Generally speaking, steady forced oscillations in linear stationary systems are always frequency- and phase-synchronized with the external excitation signal. The parametric resonance oscillations in the linear periodically nonstationary systems are always synchronized with the parameter signal variations. But to be precise, in the latter case, the synchronization and excitation can take place at the multiple frequencies of the parameter variation. The synchronization in these systems is attained "automatically". The various synchronization processes in nonlinear systems are more complicated and are the main focus of the chapter.

© Springer International Publishing AG, part of Springer Nature 2017
L. Chechurin and S. Chechurin, *Physical Fundamentals of Oscillations*,
https://doi.org/10.1007/978-3-319-75154-2_9

9.1   Oscillation Entrainment and Synchronization

Let us consider again the forced oscillations of a nonlinear system described by Eq. (7.42). Let the autonomous mode system (the right side of the equation is zero) has self-excited oscillations, which means that the solution to frequency Eq. (7.32) is found in the form of amplitude A_s and frequency ω_s of the self-excited oscillations. In the presence of the external (input) harmonic signal with the amplitude A_{in} and a frequency ω, the forced oscillations with an amplitude A can emerge at the same frequency in the system. Forced oscillation existence condition (7.42) is known as

$$\frac{A_{in}}{A}e^{-j\phi} = W^{-1}(j\omega) + W_1(A). \tag{9.1}$$

Let us also assume condition (9.1) is met and steady forced oscillations have been excited. We recall the fact that the solution to the self-oscillation existence equation, or the cross-point of the frequency response and the harmonic linearization factor, is at the same time the first parametric resonance excitation boundary. As soon as forced oscillations with another amplitude and frequency are excited in a nonlinear system, the excitation condition of parametric resonance or self-excited oscillations is no longer satisfied since the variation law of a parameter or periodic derivative is changed, and the self-excited oscillations may disappear. This phenomenon, when the self-oscillations in a nonlinear system are replaced by the forced oscillations, is called *oscillation entrainment*. The entrainment takes place in a certain range of the frequencies and amplitudes of an input signal. With a fixed frequency, the minimum amplitude of an external input signal at which the locking exists is called the entrainment threshold.

Figure 9.1 illustrates the solution of condition (9.1) in the frequency plane. For simplicity, the harmonic linearization factor of a nonlinear element $W_1(A)$ is assumed to be real. The circle with radius A_{in}/A is drawn with respect to a certain fixed point ω of the frequency response from the neighboring range of self-excited oscillations point (A_s, ω_s).

Fig. 9.1 Entrainment effect illustration

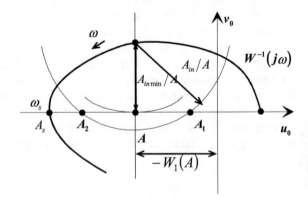

The cross-points of the circle with the real axis (or, in this case, harmonic linearization factor plot) give possible amplitudes (A_1, A_2) of the forced oscillations in the nonlinear system. Clearly, the synchronization is feasible solely at the self-excited oscillation frequency as long as there is a zero input amplitude. The minimum radius of the circle tangential to the real axis sets a certain value A and the entrainment threshold $A_{in\,min}$. Very simple conditions that define the entrainment threshold follow from Fig. 9.1:

$$|v_0(\omega)| = |\mathrm{Im}\,W^{-1}(j\omega)| = A_{in\,min}/A \qquad (9.2)$$

$$u_0(\omega) = \mathrm{Re}\,W^{-1}(j\omega) = -W_1\,(A)\,. \qquad (9.3)$$

Strictly speaking, the minimum circle radius does not always correspond to the minimum entrainment amplitude as it can be, for instance, in the case of $dA_{in}/dA <$ 0. But in the latter case, the forced oscillations are unstable although.

It is worth mentioning that synchronization takes the form of entrainment when the external signal frequency is near the frequency of self-oscillation. When the frequency difference is much bigger (several times), as the input signal amplitude grows, the so-called self-oscillation synchronization at the multiple frequencies can be observed even prior to the entrainment effect. The system oscillations frequency turns out to be the multiple of the input signal periodic variations and in-phase. The minimum input signal amplitude at which this type of synchronization takes place will be called a multiple synchronization threshold.

9.2 Low-Frequency Synchronization

Let the periodic low-frequency input signal $\omega_{in} < \omega_s$ be fed to a self-oscillatory system. Assuming as before that upper harmonics of frequencies $m\omega_s$, $m = 2, 3, \ldots$ are filtered by the linear part of the system, the low-frequency synchronization is considered with respect to two possible situations: when the self-excited oscillation component bigger or smaller than the forced oscillation component at the input of the nonlinearity. Restricting our analysis by the synchronization on the basis of a single-frequency parametric resonance, the cases lead to the synchronizations on the harmonics of either the input signal (external driving disturbance) or the forced oscillations. Let us consider these situations separately.

1. **Input harmonic synchronization**. Let us consider the self-oscillatory system with the block diagram from Fig. 9.2.

We assume that the low-frequency periodic signal $x_{in}(t)$ consisting of known harmonic components enters the system input in the form

$$x_\alpha \sin \alpha\omega_{in}t, \quad \alpha = 1, 2, \ldots, \qquad (9.4)$$

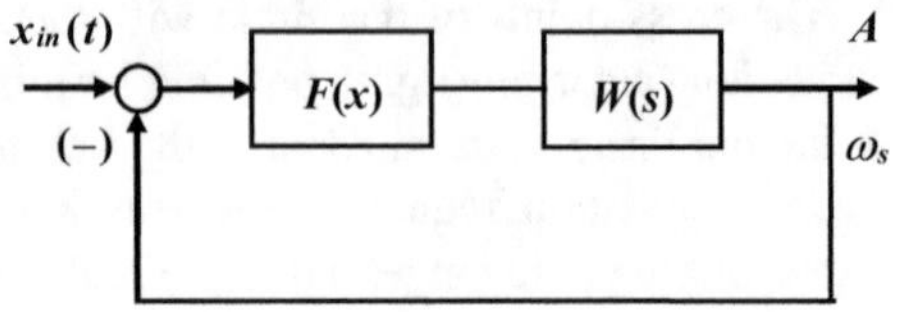

Fig. 9.2 Self-oscillatory
external signal system

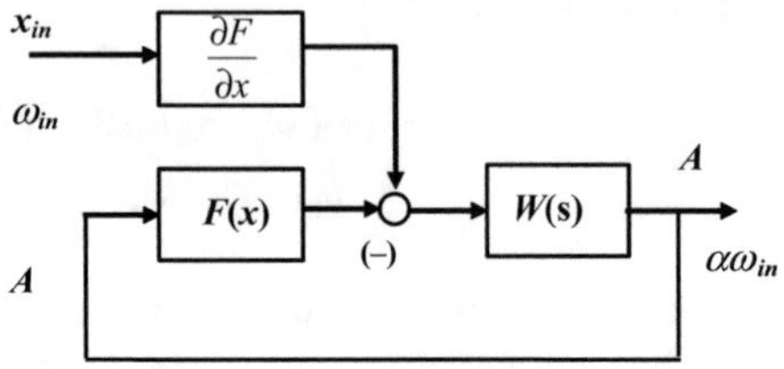

Fig. 9.3 Equivalent
representation of nonlinear
system

where ω_{in} is the first harmonic frequency of the input signal. Let us determine out of the set of harmonics (9.4) one with the closest to self-excited system oscillation frequency, i.e., $\alpha\omega_{in} \approx \omega_s$. Then, the integer value α is the frequency multiplicity. To find the entrainment threshold at that harmonic, the latter is to be derived from the system's linear part input. Two signals are the nonlinear element input. One is the signal of self-excited oscillations and another is the input signal. We assume the input amplitude x_α to have been big enough to skip the entrainment threshold, so the input signal synchronization takes place in the system. In that case, in compliance with Sect. 8.4, the system of Fig. 9.2 can be represented by the structure depicted in Fig. 9.3.

When $F(x)$ is a symmetric function, the forced oscillations with a frequency $\alpha\omega_{in}$ result in the twice frequent parameter variation of the derivative. Thus, the parameter gain with respect to the input harmonic component (9.4) is the first parametric resonance circle. Taking into account the constant parameter component, the transfer gain $W(j\varphi)$ is the ordinary circle (4.13) whose parameters are expressed in terms of harmonic linearization factors (8.19).

Thus, if the signal modulus $|x\alpha W(j\varphi)|$ is equal or exceeds the entrainment threshold with any of φ at the frequency $\alpha\omega_{in}$, that is, close to the self-excited oscillations, the input signal on the initial low frequency, ω_{in}, and the forced multiple frequency oscillations in the system become synchronized.

The synchronization conditions can be given the same as the conditions (9.2) and (9.3) shape

$$|v_0(\alpha\omega_{in})| \leq |x_\alpha W(j\varphi)| / A, \quad u_0(\alpha\omega_{in}) = -W_1(A). \tag{9.5}$$

Since the parametric resonance circle crosses the point (A_s, ω_s), as shown in Fig. 9.17, the maximum circle modulus value is the same as the modulus of that point. In the case the synchronization threshold is

Fig. 9.4 A representation of nonlinear element system

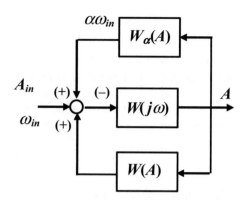

$$|x_{\alpha min}| \geq \frac{A\,|v_0(\alpha\omega_{in})|}{|W_1(A_s)|} = \frac{A\,|v_0(\alpha\omega_{in})|}{|W(j\omega_s)|} = \frac{A\,|v_0(\alpha\omega_{in})|}{|u_0(j\omega_s)|}. \tag{9.6}$$

2. **Synchronization on forced oscillation harmonics.** We turn now to the second case of the low-frequency synchronization, where the forced oscillation component at the nonlinear element input exceeds that of the self-excited oscillations.

Let the oscillation entrainment take place at a low frequency ω_{in} of a substantially big mono-harmonic input signal in the self-oscillatory system (see Fig. 9.9). This means that conditions (9.2) and (9.3) are satisfied and the forced oscillation amplitude A at the nonlinear element input can be obtained. We distinguish a nonlinearity output's harmonic which frequency $\alpha\omega_{in}$, $\alpha = 3, 5, 7, \ldots$ is the closest to the self-excited oscillation's frequency. With respect to this very harmonic, the nonlinear element is represented as a parameter varying with a period of $T = 2\pi/\omega_{in}$. The system shown in Fig. 9.9 is now transformed into the block diagram of Fig. 9.4.

For small harmonic's amplitudes, the approximation of steady synchronized steady-state oscillations at the linear part output can be written in the form

$$|x_\alpha| \approx |AW_\alpha(A)W(j\alpha\omega_{in})|. \tag{9.7}$$

The refined assessment can be derived in the form of the steady forced oscillations of the periodically nonstationary system shown in Fig. 9.4. Since the parameter varies synchronously and in-phase with the forced oscillations, its transfer function at the frequency $\alpha\omega_{in}$ looks like

$$W_\alpha(j\varphi)|_{\varphi=0} = a_0 - \rho_\alpha e^{j\alpha\varphi}, \tag{9.8}$$

where the parameter expansion coefficients ρ_α are found according to Appendix (see (17.33) and (17.35) with $m = 2\alpha \pm 1$, $\alpha = 1, 3, 5, \ldots$). It follows from the block diagram in Fig. 9.4 that

$$|x_\alpha| \approx \left| AW_\alpha(A) \frac{W(j\alpha\omega_{in})}{1 + W(j\alpha\omega_{in}) \, W_\alpha(j\varphi)|_{\varphi=0}} \right|. \tag{9.9}$$

Finally, it should be noted that the parametric resonance of the parameter variation frequency $2\alpha\omega_{in}$ can take place in the system of Fig. 9.4. Then, the parametric oscillations excited at the frequency $2\alpha\omega_{in}$ are also synchronized with the low frequency of the input. The excitation conditions are known to be met provided that the inverse frequency characteristic point $\alpha\omega_{in}$ gets inside the circle

$$W_\alpha(j\varphi) = a_0 - \frac{a_\alpha}{2j} e^{-j\varphi} \tag{9.10}$$

As a matter of fact, this case of low-frequency synchronization is not synchronization at all, but the high-frequency system entrainment by a low-frequency signal. The system oscillations are excited at the frequency of the higher harmonics of the forced oscillations.

9.3 High-Frequency Synchronization

Let us now assume that the oscillations with a frequency ω_1, that is, close to the self-excited oscillation frequency $\omega_1 \approx \omega_s$, have become stable in a nonlinear self-oscillatory system as a result of synchronization at the input frequency of $n\omega_1, n = 3, 5, \ldots$. In other words, let us assume that the two-frequency motion $x = x_1 + x_n$ has become stable in the system, and the motion is described by Eq. (7.42):

$$G(s)(x_1 + x_n) + H(s)F(x_1 + x_n) = H(s)x_{in}(t). \tag{9.11}$$

Assuming the high-frequency oscillations to be small, let us approximate the (9.11) by

$$G(s)(x_1 + x_n) + H(s)\left[F(x_1) + \frac{\partial F(x_1)}{\partial x} x_n \right] \approx H(s)x_{in}(t). \tag{9.12}$$

Equation (9.12) can be given the operator form and employ the harmonic linearization factors with respect to the first and n th harmonics:

$$G(s)(x_1 + x_n) + H(s)\left[W_1 x_1 + W_n x_1 + \frac{\partial F(x_1)}{\partial x} x_n \right] \approx H(s)x_{in}(t). \tag{9.13}$$

We group the members of the same frequencies now

$$G(j\omega_1)x_1 + H(j\omega_1)\left[W_1(A)x_1 + \frac{\partial F(x_1)}{\partial x} x_n \right] = 0 \tag{9.14}$$

$$G(jn\omega_1)x_n + H(jn\omega_1)[W_n(A)x_1] = H(jn\omega_1)x_{in}. \tag{9.15}$$

The high-frequency component is derived from Eq. (9.15) as

$$[x_{in} - x_1 W_n(A)]W(jn\omega_1) = x_n \tag{9.16}$$

and substituted into Eq. (9.15):

$$G(j\omega_1)x_1 + H(j\omega_1)\left\{ W_1(A)x_1 + \frac{\partial F(x_1)}{\partial x}[x_{in}W(jn\omega_1) - x_1 W_n(A)W(jn\omega_1)]\right\} = 0. \tag{9.17}$$

Let us pay attention to the second item in the squiggle brackets. The derivative of the periodic argument of a nonlinear function is a periodic parameter. The first item in the square brackets includes the input harmonic with an arbitrary time shift τ, or phase shift φ relative to the parameter variations, whereas the second item cannot have the shift, i.e., $\varphi = 0$ for the second item. Having said this, we need to use different parameter transfer functions for the element in the square brackets, so Eq. (9.17) takes the form

$$x_1 + W(j\omega_1)\left\lfloor W_1(A)x_1 + W(j\varphi)x_{in}W(jn\omega_1) - W(j\varphi)|_{\varphi=0}x_1 W_n(A)W(jn\omega_1)\right\rfloor = 0. \tag{9.18}$$

That results in the following ratio:

$$\frac{x_{in}}{x_1} = W^{-1}(j\varphi)\left\{ W^{-1}(jn\omega_1)\left[W^{-1}(j\omega_1) + W_1(A)\right] - W(j\varphi)|_{\varphi=0}W_n(A)\right\} = 0. \tag{9.19}$$

The same equation can be derived from the structure representation of the system in Fig. 9.5. According to the definitions (9.1)–(9.3), the expression in the square brackets is the entrainment threshold D of the system at the frequency ω_1, as it is shown in Fig. 9.6. Thus, Eq. (9.19) is rewritten as follows:

$$\frac{x_{in}}{x_1} = W^{-1}(j\varphi)\left[DW^{-1}(jn\omega_1) - W(j\varphi)|_{\varphi=0}W_n(A)\right]. \tag{9.20}$$

The parameter transfer gain $W(j\varphi)$ in Eq. (9.20) is generally the coupling coefficient for harmonics (17.23). If $m = 1$

$$a_{1n} = W(j\varphi) = -\frac{1}{2j}\left(a_{\frac{n-1}{2}}e^{j\frac{n-1}{2}\varphi} + a_{\frac{n+1}{2}}e^{-j\frac{n+1}{2}\varphi}\right) \tag{9.21}$$

and correspondingly

$$W(j0) = -\frac{1}{2j}\left(a_{\frac{n-1}{2}} + a_{\frac{n+1}{2}}\right). \tag{9.22}$$

If we migrate to modulus equality in (9.20):

$$\frac{A_{in}}{A} = \left|\frac{DW^{-1}(jn\omega_1)}{W(j\varphi)} - \frac{W(j\varphi)|_{\varphi=0}W_n(A)}{W(j\varphi)}\right|. \tag{9.23}$$

Fig. 9.5 Block diagram of
definition (9.19)

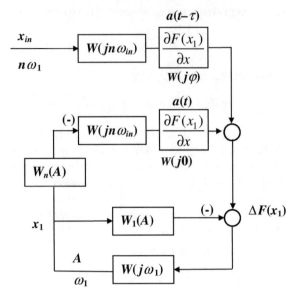

Fig. 9.6 Synchronization
condition evaluation

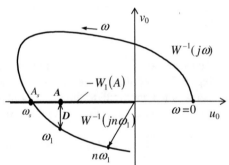

The radius of the second central circle in relation (9.21) is less than that to the first circle usually. Therefore, by neglecting the coefficient $a_{(n+1)/2}$, the synchronization threshold assessment can be derived from equality (9.23) as

$$\frac{A_{in}}{A} = \frac{a_{(n-1)/2}}{2} \left| DW^{-1}(jn\omega_1) - W_n(A) \right|. \tag{9.24}$$

Here, the coefficients $a_{(n\pm1)/2}$ are expressed in terms of the harmonic linearization factors (17.33) and (17.34). Thus, expression (9.24) provides the evaluation of the high-frequency synchronization threshold at the given frequency ω_1 that is supposed to be close to the self-oscillation frequency, and the amplitude A, that is defined by the entrainment threshold at this.

9.4 Parametric Synchronization

The synchronization was achieved in the previous problems by means of an external synchronizing signal. Let us consider the synchronizing problem of a nonlinear oscillatory system with the help of a periodically variable parameter which in turn is controlled by an external harmonic signal. Out of numerous engineering applications of the problem, we exemplify an self-oscillatory system in which the nonlinearity is in parallel connection with the periodically controlled symmetric parameter, as shown in Fig. 9.7.

We assume that the periodic parameter variation

$$a(t - \tau) = A_{in} \sin 2\omega_{in}(t - \tau),$$

excites the system oscillations

$$x(t) = A \sin \omega_{in} t$$

synchronized with the parameter oscillations. We also assume that these oscillations are stable, and their frequency is close to that of the self-oscillations. The smallest required amplitude A_{in} of the synchronizing signal is defined, as before, by the entrainment threshold at the frequency ω_{in}. Equation (9.1) can be used to evaluate the entrainment threshold D (see Fig. 9.6), that is,

$$A_{in \, min} = A|W^{-1}(j\omega_{in}) + W_1(A)| = AD. \tag{9.25}$$

On the other hand, according to Chap. 2, the synchronizing signal is the parameter output signal

Fig. 9.7 Parametric
synchronization diagram

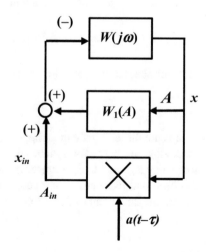

Fig. 9.8 On synchronizing
parameter evaluation

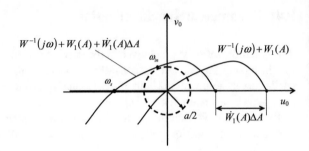

$$x_{in}(t) = W(j\varphi)x(t) = -\frac{a}{2j}e^{-j\varphi}x(t). \tag{9.26}$$

It follows from the modulus equality that

$$A_{in} = aA/2. \tag{9.27}$$

Finally, taking into account formulation (9.26), the estimation is obtained as

$$a \geq 2\left|W^{-1}(j\omega_{in}) + W_1(A)\right| = 2D. \tag{9.28}$$

Another evaluation of the synchronizing parameter can be derived in the plane of
the inverse amplitude–frequency characteristic shown in Fig. 9.8.

The amplitude–frequency characteristic which modulus is used in the estimation
(9.28) is the inverse frequency characteristic shifted to the right by the quantity of
$W_1(A)$. The shift is $W_1(A_a)$ and the characteristic crosses the origin, the latter is
the point of self-oscillations, as shown in Fig. 9.8. The shifted amplitude–frequency
characteristic within the small neighborhood of the origin under a small deviation
ΔA of the self-oscillation A_s is about

$$W_{rep}^{-1}(\omega, A_s + \Delta A) \approx W^{-1}(j\omega) + W_1(A_s) + \frac{\partial W_1(A_s)}{\partial A}\Delta A \tag{9.29}$$

in which the modulus of shifting to either the right or the left is equal to

$$|\Delta u_0| = \left|\dot{W}_1(A_s)\Delta A\right|. \tag{9.30}$$

It is not difficult to define graphically the point of contact of the central circle and
the hodograph W_{rep}^{-1}. The synchronizing threshold with respect to the synchronizing
frequency ω_{in}, and the parameter amplitude a (that is equal to the circle diameter)
can be found like that. In any case, the synchronizing parameter threshold cannot
exceed the increment (9.30). From which "a vulgar" approximation follows

$$a \leq \left|2\dot{W}_1(A_s)\Delta A\right|. \tag{9.31}$$

Other concepts of parametric synchronization are also possible.

9.5 Synchronization Condition Evaluation Examples[1]

1. Self-oscillation finding. Let us consider the self-excitatory system presented in Fig. 9.9. Let its linear subsystem be described by the equation

$$W(j\omega) = \frac{1}{(j\omega)^3 + 2(j\omega)^2 + j\omega}$$

and the nonlinear element of a feedback loop has the characteristic

$$F(x) = \mathrm{sgn}x.$$

The frequency ω_s and amplitude A_s of the self-excited oscillations in the system are found using the harmonic linearization factor (7.21) and frequency equality (7.32) in the form

$$\frac{1}{(j\omega)^3 + 2(j\omega)^2 + j\omega} = -\frac{\pi A}{4},$$

from we obtain $\omega_s = 1$ and $A_s = A/\pi$.

The same values can be obtained graphically from the location of the contact point of the inverse Nyquist hodograph and the negative complex harmonic linearization factor plot (see Fig. 9.10).

2. Entrainment threshold determination. We are interested to find the minimum input signal variation amplitude $A_{in\,min}$ of frequency $\omega_{in} = 0.7$ rad/s at which the oscillation entrainment happens. Then, the entrainment conditions can be formulated in similar to (9.2) and (9.3) manner

$$|v_0(\omega)| = \left|\mathrm{Im}W^{-1}(j\omega)\right| = A_{in\,min}/A = 0.357$$
$$u_0(\omega) = \mathrm{Re}W^{-1}(j\omega) = -W_1(A) = 0.98,$$

Fig. 9.9 Self-oscillatory system under investigation

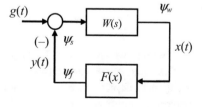

---

[1]The paragraph is presented by V. V. Semenov, see also [11].

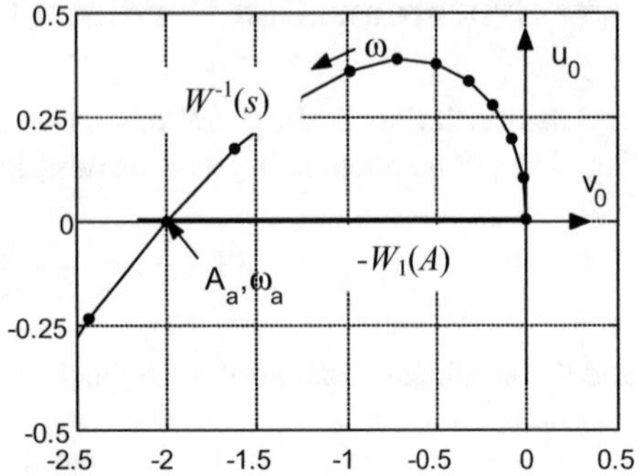

Fig. 9.10 Self-excited
oscillation evaluation

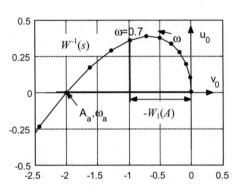

Fig. 9.11 On evaluation of
entrainment conditions at
frequency 0.7 rad/s

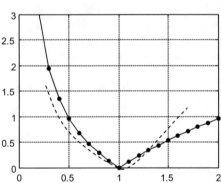

Fig. 9.12 Entrainment
threshold versus frequency.
Solid line: calculation, dash
line: experiment

from which we obtain $A \approx 1.247$, $A_{in\ min} \approx 0.445$.

The same values can be found geometrically, as shown in Fig. 9.11. A solid line
in Fig. 9.12 depicts the calculated boundary of the synchronization entrainment. The
numerical experiment scheme and the experiment results are compared in Figs. 9.13
and 9.14, correspondingly.

3. Low-frequency synchronization. Let us derive the low-frequency synchronization conditions for the system from the first example of this paragraph, Fig. 9.9.

Fig. 9.13 Block diagram of
the numerical experiment

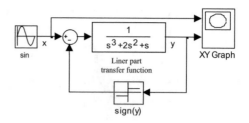

Fig. 9.14 Oscillation
entrainment. $A = 0.43$,
$\omega = 0.7$ rad/s

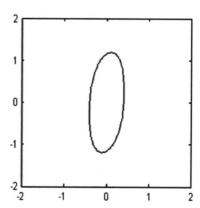

As noted above, we distinguish the two cases of low-frequency synchronization. In
the first case, the entrainment appears since there are high-order harmonics in the
low-frequency input signal. Let the input signal be

$$x_{in} = \alpha_1 \cdot \sin(\omega t) + \alpha_2 \cdot \sin(2\omega t),$$

where $\alpha_1 \cdot$ and $\alpha_2 \cdot$ are the amplitudes of the first and second harmonics, correspondingly.

Let us find the minimum amplitude of the second harmonic under the low-
frequency synchronization condition provided that there is no entrainment at the
first harmonic. According to conditions (9.2) and (9.3) at the input frequency
$\omega = 0.45$ rad/s,

$$|v_0(\omega)| = \left|\mathrm{Im}W^{-1}(j\omega)\right| = A_{in\,\min}/A = 0.357$$
$$u_0(\omega) = \mathrm{Re}W^{-1}(j\omega) = -W_1(A) = 0.405,$$

and the entrainment threshold at the first harmonic is

$$A \approx 3.144, \quad A_{in\,\min} = 1.129.$$

The first harmonic amplitude is to be chosen lower than the locking threshold, so
let us take $A_{in\,\min} = 0.5$.

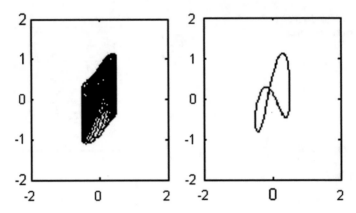

Fig. 9.15 Simulation results

According to (9.6), the minimum second harmonic amplitude can be defined as

$$x_{2\min} \geq \frac{A\,|v_0(2\omega_{in})|}{|u_0(j\omega_a)|} = \frac{0.786 \cdot \left|\mathrm{Im}(W^{-1}(j\varphi)|_{\omega=2\omega_{in}}\right|}{\left|\mathrm{Re}(W^{-1}(j\varphi||_{\omega=\omega_a})\right|} = \frac{A \cdot 0.171}{2},$$

where A is the steady-state oscillation amplitude in the system calculated by the second part of Eq. (9.6), i.e.,

$$A = \frac{4}{\pi\,\left|\mathrm{Re}(W^{-1}(j\omega)|_{\omega=2\omega_{in}})\right|} = 0.786.$$

So the minimum amplitude of system synchronization at the input harmonic is

$$x_{2\min} = \frac{0.786 \cdot 0.171}{2} = 0.0672.$$

Figure 9.15 presents the simulation results of synchronizing process. The left picture reflects the condition when the second harmonic amplitude 0.03 is not enough for the synchronizing. The right picture illustrates the low-frequency synchronization with $x_{2\min} = 0.07$.

In the case of the synchronizing at the self-oscillation harmonics, when the forced component at the nonlinear element input exceeds the natural one, the synchronization occurrence condition is described by expression (9.9) as long as the forced oscillation component. We assume now that the system's input is the oscillatory signal of frequency $\omega_{in} = 0.3$ rad/s and high enough amplitude for the low-frequency entrainment at ω_{in}; in other words, conditions (9.2) and (9.3) hold. Let us evaluate the forced oscillation amplitude A

$$|v_0(\omega)| = \left|\mathrm{Im}\,W^{-1}(j\omega)\right| = A_{in\,\min}/A = 0.273.$$
$$u_0(\omega) = \mathrm{Re}\,W^{-1}(j\omega) = -W_1(A) = -0.18.$$

From which it follows that $A \approx 7.073$, $A_{in\,\min} = 1.93$.

It is obvious that the third harmonic frequency is the closest to the self-oscillation frequency in the system. The harmonic linearization factor to the nonlinear subsystem with respect to the third harmonic is

$$W_3(A) = \frac{j}{\pi A} \int\limits_0^{2\pi} F(A \sin \psi) e^{-j3\psi}\, d\psi = \frac{4}{3\pi A}.$$

According to (9.8), the parameter transfer gain at the frequency $\omega = 3\omega_{in}$ becomes

$$W_\alpha(j\varphi)\big|_{\varphi=0} = a_0 - a_\alpha/2j,$$

where

$$a_0 = \frac{1}{2\pi} \int\limits_0^{2\pi} \frac{dF(A \sin \psi)}{dx}\, d\psi = 0,$$

$$\frac{a_3}{2j} = 3W_3(A) + \frac{A}{2}\frac{dW_3(A)}{dA} = \frac{10}{3\pi A}.$$

Substituting a_0, a_3, we arrive at

$$W_3(j\varphi)\big|_{\varphi=0} = -\frac{10}{3\pi A}.$$

Based on these results, the third harmonic amplitude can be evaluated as

$$|x| \approx \left| A W_3(A)\frac{W(j3\omega_{in})}{1 + W(j3\omega_{in})W_3\,(j\varphi)\big|_{\varphi=0}} \right| = 0.287.$$

Thus, the third harmonic amplitude has to be not more than 0.287 to observe the forced oscillation harmonic synchronization.

Example 3. Let the oscillatory signal of frequency $3\omega_{in} = 3.3$ rad/s enter the system as shown in Fig. 9.9. The input minimum amplitude can be evaluated by the expression (9.24) as

$$A_{in} = A\frac{a_3}{2}\left|DW^{-1}(jn\omega_1) - W_n(A)\right|,$$

where $D = \mathrm{Im}(W(j\omega_{in})) = 0.231$ and

Fig. 9.16 Example. On high-frequency synchronization condition analysis

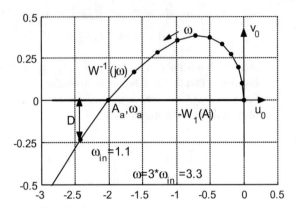

Fig. 9.17 Simulation results. Left: $A_{in} < 2.6$, no synchronization; right: $A_{in} > 2.6$, synchronization

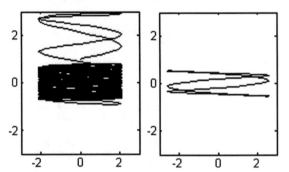

$$W_3(A) = \frac{j}{\pi A} \int_0^{2\pi} F(A \sin \psi) e^{-j3\psi} d\psi = \frac{4}{3\pi A}.$$

Substituting the results to evaluation (9.24), we can derive $A_{in\,min}$ as

$$A_{in\,min} = A \frac{a_3}{2j} \left| DW^{-1}(j3\omega_1) - W_3(A) \right| \approx 2.61.$$

Figure 9.16 gives the graphical solution of the problem in the plane of the inverse amplitude–phase–frequency characteristic $W^{-1}(j\omega)$.

We illustrate the analytics by a numerical experiment. The setting is given in Fig. 9.13. The input synchronizing oscillatory signal frequency is set to be 3.3 rad/s. The high-frequency entrainment threshold is found by varying the input signal amplitude within a certain range. Figures 9.17 and 9.18 depict the simulation results.

Example 4. Let us estimate the synchronization threshold for parametric synchronization. The synchronizing signal is assumed to be $a(t - \tau) = A_{in} \sin 2\omega_{in}(t - \tau)$ at $\varpi_{in} = 0.8$. Then, the entrainment threshold can be found according to condition (9.2) as

Fig. 9.18 Oscillograms of signals. x—input, y—forced oscillations

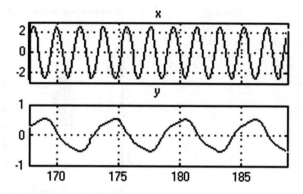

Fig. 9.19 Numerical experiment setting for parametric synchronization study

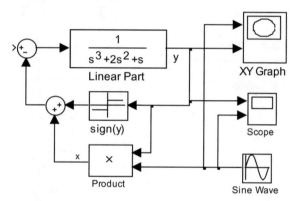

$$D = \mathrm{Im}(W(j\omega_{in})) = 0.288$$

and the synchronization threshold has the following value according to (9.28):

$$A_{in\,\min} = 2 \cdot D = 0.576.$$

Figure 9.19 presents one of the possible block diagrams for a numerical experiment. The simulation results are illustrated in Figs. 9.20 and 9.21.

Fig. 9.20 Simulation
results. Left: $A < 0.55$, no
synchronization; right:
$A > 0.55$, synchronization

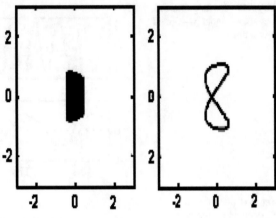

Fig. 9.21 Parametric
synchronization boundary.
Solid line: predicted, dashed
line: experiment

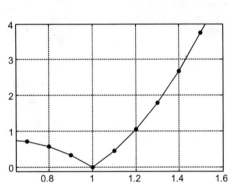

Part IV
Parametric Oscillation Control

This part adds to the analysis of parametric damping and exciting of spring pendulum oscillations. We demonstrate how changes in parameter variation phase and magnitude result in growing or vanishing of output oscillations. This leads to the same idea that drives the automatic control design to improve the output process quality by a feedback with controllable periodic parameter variations—in other words, parametric regulator. We provide various applicable schemes of parametric control and coordinate control and compare them with several examples.

Chapter 10
Parametric Damping and Exciting of Oscillations

10.1 Parametric Damping of Spring Pendulum Oscillation

Let us consider an oscillatory system, comprising the gravitational pendulum on an elastic suspension (spring). Thus, the vertical motion of the mass happens when the spring is compressed or elongated. Since a parameter (length) of the gravitational pendulum sways periodically varies, we can sometimes observe the parametric oscillations of this angular coordinate.

The simplified description of this oscillatory system is taken in the form

$$\ddot{x} + \zeta\dot{x} + \omega_x^2 x = l_0\dot{\alpha}^2 + f(t)$$
$$\ddot{\alpha} + [l_0^{-2} - 2l_0^{-3}\Delta l(t)]\,\zeta\dot{\alpha} + [l_0^{-1} - \ell_0^{-2}\Delta l(t)]\,g\alpha = 0. \tag{10.1}$$

Here, $x(t) = \Delta l(t)$ are vertical oscillations of the mass relative to its unperturbed position, $\alpha(t)$ are the angular mass sways, $\omega_x^2 = c/m$, $\omega_\alpha^2 = g/l_0$ are the natural frequencies of the vertical oscillations and the angular swings, correspondingly, c is spring stiffness, m is mass, l_0 is the suspension length in the unperturbed state, $f(t)$ is an external force generator signal along the vertical axis, and ζ is a damping factor. We also assume the conditions $x(t) = \Delta l(t) < l_0$, $\sin\alpha \approx \alpha$, $\ddot{x} < g$ are held.

The first of Eq. (10.1) describes the nonlinear relation between the angular swings and the parameter oscillations. The second of the equations provides an angular swing profile. Thus, the spring pendulum is a nonlinear periodically nonstationary system with a controlled parameter or length. In a single-frequency harmonic approximation, the periodic solution of angular swings is searched in the form

$$\alpha(t) = A\sin\omega_\alpha t.$$

In that case, the nonlinear term in Eq. (10.1) can be transformed as follows:

$$l_0\dot{\alpha}^2 = l_0\left[\frac{d}{dt}(A\sin\omega_\alpha t)\right]^2 = A^2 l_0\omega_\alpha^2\cos^2\omega_\alpha t = 0.5A^2 l_0\omega_\alpha^2(1 + \cos 2\omega_\alpha t)$$

© Springer International Publishing AG, part of Springer Nature 2017
L. Chechurin and S. Chechurin, *Physical Fundamentals of Oscillations*,
https://doi.org/10.1007/978-3-319-75154-2_10

and the parameter l_0 gains an additional constant component $x_0 = 0.5A^2l_0\omega_\alpha^2$ and the variable component of a doubled frequency

$$\tilde{x} = 0.5A^2l_0\omega_\alpha^2|W_x(j\omega_x)|\cos(\omega_x t + \psi_x) = 0.5A^2l_0\omega_\alpha^2 |W_x(j\omega_x)| \sin \omega_x t,$$

where

$$W_x(j\omega_x) = \frac{1}{s^2 + \zeta s + \omega_x^2} \Big|_{s=j\omega_x} \tag{10.2}$$

is the frequency response for the vertical oscillations, and $\psi_x = \arg W_x(j\omega_x) = -\pi/2$ is its phase response.

Within the chosen notation, the output harmonic signal of the controlled parameter is as

$$\begin{aligned} x_{out}(t) &= \tilde{x}(t)\alpha(t) \\ &= 0.5A^3l_0\omega_\alpha^2 |W_x(j2\omega_\alpha)| \sin 2\omega_\alpha t \sin \omega_\alpha t \\ &\approx 0.25A^3\omega_\alpha^2 l_0 |W_x(j2\omega_\alpha| \cos(\omega_\alpha t), \end{aligned}$$

where $\omega_x = 2\omega_\alpha$.

Employing the operator form in the angular sway equation and moving the terms containing $\Delta l(t)$ to the right side, the transfer function $W_\alpha(s)$ of the angular swings takes the form

$$W_\alpha(s) = \frac{\alpha(s)}{\Delta l(s)} = \frac{2\zeta s + 1}{s^2 + l_0^{-2}\zeta s + \omega_\alpha^2}, \tag{10.3}$$

and its frequency response is $W_\alpha(j\omega_\alpha)$, where $\psi_\alpha = \arg W_\alpha(j\omega_\alpha) = -\pi/2$.

Now, the angular swings at the system output take the form

$$\alpha(t) = x_{out}(t)W_\alpha(j\omega_\alpha) = 0.25A^3\omega_\alpha^2 l_0|W_x(j2\omega_\alpha)||W_\alpha(j\omega_\alpha)| \sin(\omega_\alpha t).$$

Dividing the last expression by $\alpha(t) = A \sin \omega_\alpha t$, as assumed before, the real gain of an open-loop system is found as

$$W_{ol}(A, \omega_\alpha) = 0.25A^2\omega_\alpha^2 l_0|W_x(j2\omega_\alpha^2)||W_\alpha(j\omega_\alpha)|.$$

Since the frequency equation of a negative feedback system

$$W_{ol}(A, \omega_\alpha) + 1 = 0$$

has no solution for $\ell_0 > 0$, there is no parametric resonance in the gravitation spring pendulum within the first harmonic approximation.

Figure 10.1 displays the spring pendulum model with $m = 1$ kg, $l_0 = 1$ m, $c = 40$ N/m, and $\zeta = 0.1$.

Disconnecting the parameter controlling loop from the feedback, we arrive at the single circuit model of a variable length pendulum. The latter is the linear system with independent periodic length (parameter) variation. The system can have two statuses: (1) a zero stable equilibrium and (2) the unlimited growth of parametric resonance oscillations when the amplitude of the generator *Sine Wave* is $A_{sw} > 1.5$. But the examination of the spring pendulum model (see Fig. 10.1) reveals the ranges of steady forced parametric oscillations as well as monotonously divergent output at $A_{sw} > 14$ (due to the growth of the constant component of the parameter variation) besides the zero stable equilibrium at the *Sine Wave* amplitude $A_{sw} > 1.5$. Although $A_{sw} > 14$ far exceeds the model validity assumptions. **It is significant that the parametric control eliminates the self-excitation of parametric resonance in the system and introduces the damping and the threshold to steady parametrically forced oscillations in the gravitation spring pendulum system.**

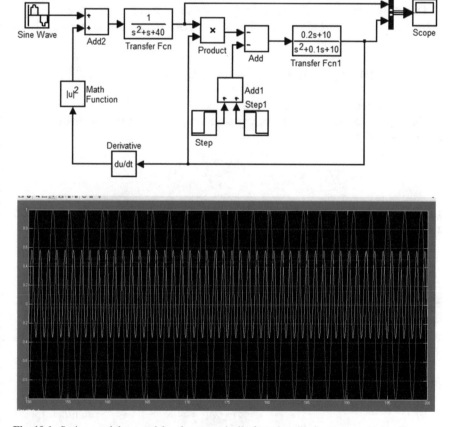

Fig. 10.1 Spring pendulum model and parametrically forced oscillation process ($A_{sw} > 5$)

10.2 Parametric Excitation of Gravitation Spring Pendulum Oscillations

Double spring pendulum. The simplified description of a double mathematical pendulum has the form

$$(m_1 + m_2)\ell_1\ddot{\varphi}_1 + m_2\ell_2\ddot{\varphi}_2 + (m_1 + m_2)g\varphi_1 = 0$$
$$\ell_1\ddot{\varphi}_1 + \ell_2\ddot{\varphi}_2 + g\varphi_2 = 0. \tag{10.4}$$

In the agreed notations, $\varphi_1, \varphi_2; m_1, m_2; \ell_1, \ell_2$ are deviation angles (or coordinates), masses, and lengths of the upper (first) and lower (second) pendulums, respectively. The operator equation is obtained by excluding the angle φ_2 as

$$m_1\ell_2 s^4 + g(1 + \ell_2/\ell_1)s^2 + g^2/\ell_1 = 0. \tag{10.5}$$

Due to the vertical oscillations the first pendulum length varies as $\ell_1 = \ell_{01} + \Delta\ell_1(t)$, $\ell_{01} \gg \Delta\ell_1(t)$. Using the method of stationarization, we transform Eq. (10.5) to

$$m_1\ell_2 s^4 + g(1 + \ell_2/\ell_{01})s^2 + g^2/\ell_{01} = [g(\ell_2/\ell_0^2)s^2 + g^2/\ell_0^2]0.5\Delta\ell_1 e^{-j\varphi}, \tag{10.6}$$

where $\Delta\ell_1$ is the parameter variation amplitude.

The transfer function and frequency response ($s = j\omega$) of this oscillatory object come from (10.6)

$$W(s) = \frac{g(\ell_2 s^2 + g)/\ell_{01}^2}{m_1\ell_2 s^4 + g(1 + \ell_2/\ell_{01})s^2 + g^2/\ell_{01}}. \tag{10.7}$$

Let us agree on certain physical parameter values $m_1 = m_2 = 0.5$ kg and $l_1 = l_2 = 0.5$ m. Thus, the transfer function becomes

$$W(s) \cong \frac{78s^2 + 1540}{s^4 + 78s^2 + 770}.$$

Figure 10.2 depicts the double spring pendulum model and the first parametric resonance excitation process at the second natural frequency $\omega_{02} = 8$ rad/s. The left parametric control circuit is the double frequency oscillation circuit of the spring of stiffness 256 N/m that is guided by centrifugal force.

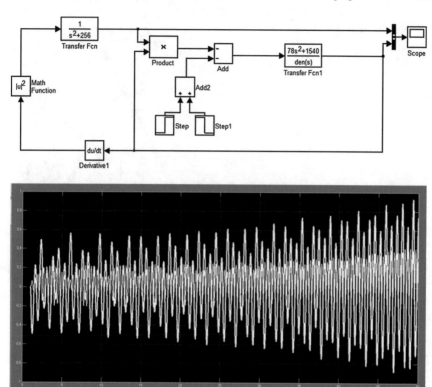

Fig. 10.2 Double spring pendulum model and parametric resonance excitation process

10.3 Physical Interpretation of Parametric Oscillation Excitation and Damping

Prior to elaborating the parametric regulators design requirements we need to discuss the physics of parametric damping and excitation. We will use simple swing as an example.

The parametric oscillations of a swing are excited due to the periodic motion of a swinging person along the suspension axis, as shown in Fig. 10.3. It can be seen as the periodic variation of inertia or the effective length of the pendulum.

The easiest and simplest way to swing is to keep the oscillation frequency as near as possible to the natural frequency of free oscillations of the swing with the motionless load. It requires the parameter variation frequency (or the vertical load/person displacement) to be two times higher than the swing oscillation frequency. (Needles to note that if two persons swing, they squat by turns to provide the twofold parameter variation over the period of swing oscillations.) The observed by this parametric resonance is called the first (main) parametric resonance. But these conditions of parametric oscillation excitation of the swing are not still sufficient. Another important

Fig. 10.3 Swing oscillations

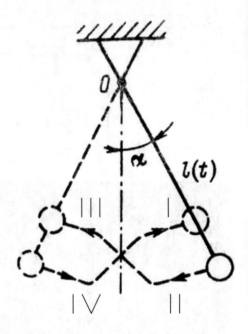

parametric excitation condition involves the parameter oscillation phase relative to the swing oscillations. The downward motion of the load, that is, the squat of the person, has to be initiated as soon as the swing reaches its maximum deflection in either direction, i.e., at zero angular velocity. In turn, the load/person upward motions, i.e., the decrease in the pendulum effective length, have to be at the zero swing deflection from the vertical position, in other words, at the maximum angular velocity.

To attain the most efficient swinging growth, we need to increase the effective pendulum's length in the second and fourth quarters (the acceleration quarters) of the pendulum oscillation period and to decrease that in the first and the third quarters (the damping quarters), as shown in Fig. 10.3. The parametric damping of the swing is achieved by the inverse: by reducing the effective length just at the maximum sway coordinate deflection in the second and the fourth quarters and by increasing the effective length when passing zero deflection point in the first and the third quarters, as shown in Fig. 10.4. This provides an extra braking torque to the sways.

This damping phenomenon takes place in the spring pendulum oscillations examined in Sect. 10.2 since the maximum velocity, centrifugal force, spring extension, and length of the pendulum are attained at the moment when the swing passes its vertical (zero coordinate) position. That is why we cannot observe the parametric resonance in the gravitational spring pendulum.

Nevertheless, why the parametric resonance oscillation emerges in double spring pendulum, which scheme is given in Fig. 10.2? The fact is that the second suspended inertia pendulum damps the first pendulum oscillations and adds certain lag to them.

Fig. 10.4 Parametric
damping of swing

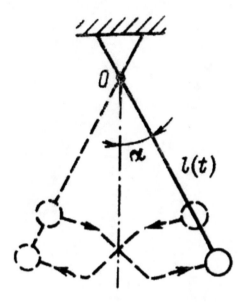

Due to the lag, the parametric damping sequence of the first pendulum oscillations
is shifted toward the parametric excitation. That cause is similar to the potentially
dangerous parametric resonance excitations which appear in moving containers with
loosely fixed cargo, liquids, and bulk cargos in partially full tankers.

Chapter 11
Parametric Regulators

11.1 On Parametric Controllability

A swinging person has been known to be able to control sway oscillations by the periodic variation of system parameter that is the effective pendulum length. It follows from Sect. 10.3 that the operator can vary not only the parameter variation amplitude but also the variation phase; thereby, the excitation and damping of the parametric swing oscillations are affected. This is the manual parametric control.

In the spring pendulum presented in Sect. 10.1, the parameter (or pendulum length) is controlled itself by the spring feedback loop. It results in the damping of the parametric resonance oscillations. This type of parametric control is called automatic. The parametric control in the double spring pendulum is also parametric (see Sect. 10.2) but it destabilizes the oscillations in the form of parametric resonance. The common disadvantage of the automatic parametric control in both examples is that it is impossible to vary or assign the parameter variation phase. It shortens the range of parametric controllability.

Figure 11.1 displays parametric oscillation excitation conditions (4.20): the inverse frequency response $W^{-1}(j\omega)$ of the oscillatory object and the parametric resonance circle $W(j\varphi)$ are plotted. The system belongs to the region of oscillation damping. The damping degree is defined by the distance between points 1 and 2. As the parameter variation amplitude $\Delta\ell$ rises, these points meet each other and the parametric resonance takes place. If the parameter variation phase is somehow moved from point 2 to point 3 providing the phase shift by the angle $\pi/2$, the damping degree increases.

11.2 General Design Principles of Parametric Regulators

Wrapping up what has been said in the previous paragraph, the parametric regulator is to hold the parameter variations at the required amplitude, phase, and frequency that is

© Springer International Publishing AG, part of Springer Nature 2017 127
L. Chechurin and S. Chechurin, *Physical Fundamentals of Oscillations*,
https://doi.org/10.1007/978-3-319-75154-2_11

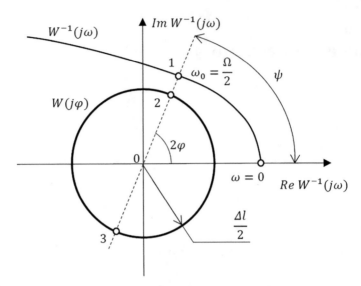

Fig. 11.1 Parametric oscillation damping degree

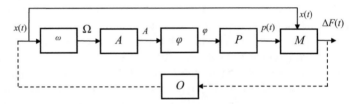

Fig. 11.2 General block scheme of parametric regulator

the doubled coordinate oscillation frequency. Figure 11.2 illustrates the general block diagram of the parametric regulator (presented first as a patent [12]) consisting of five converters: the frequency converter "ω", the amplitude converter "A", the phase converter "φ", the function convertor "P", and the modulator "M". The coordinate oscillation signal $x(t)$ enters the regulator input from the controlled object "O". The parameter variations $p(t)$ generated by the regulator are fed to the controlled object input as signal $\Delta F(t)$ from the modulator "M".

The order of the connections to the converters "ω", "A", and "φ" can vary. Some of the converters, but not the modulator, may be trimmed for specific problems. Let us have a detailed look at the problems and implementation versions of the converters.

Frequency converter. For most of applications, the frequency converter doubles the natural frequencies of the object (with respect to the first parametric resonance oscillations). In the relatively rare cases of high-order parametric resonance oscillations (the second, third, etc.), the frequency converter has to include a resonance filter for certain harmonic. But even in this case, we need to double the frequency

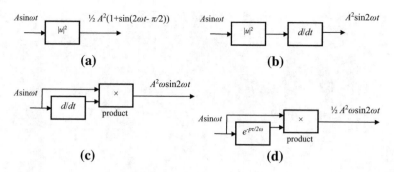

Fig. 11.3 Frequency converter versions

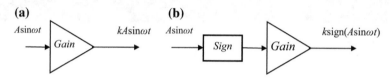

Fig. 11.4 Amplitude converters

since the nth-order parametric resonance is the first parametric resonance at the n th harmonic.

There are several methods of frequency doubling, as shown in Fig. 11.3.

All the converters double the parameter oscillation frequency and have the output oscillation amplitude proportional to the squared amplitude of coordinate oscillations. The converter shown in Fig. 11.3a adds a constant component to the parameter periodic variation. The parameter variation amplitude in the converters of Fig. 11.3b, c is proportional to the constant frequency of the coordinate oscillations.

Amplitude converter. The dependence of the parameter variation amplitude on coordinate oscillation amplitude makes no significant impact on the parametric oscillation excitation processes if the amplitudes are small. While the amplitudes are high, the dependence leads frequently to the asymptotic stability loss of a parametric control system. This dependence can be diminished or eliminated with the help of an amplitude converter. A conventional amplifier (see the *Gain* unit in Fig. 11.4) is a primary elementary amplitude converter to weaken the parameter-amplitude relationship by decreasing the amplifier gain coefficient.

Another radical approach to get rid of the parameter-coordinate oscillation amplitude dependence is the introduction of nonlinear two-position boundary controller (see the *Sign* unit in Fig. 11.4b), the level of a parameter limitation is set by the *Gain* amplifier.

Phase converter. The setting of the required parameter variation phase is performed in the range $0 < \varphi < \pi/2$ by means of phase-shifting circuits or the introduction the lag τ in the limits $0 < \tau < \pi/2\omega$ (see the *Transport Delay*). Beyond the limits, the lag effect has a periodic behavior with a period $T = \pi/2\omega$.

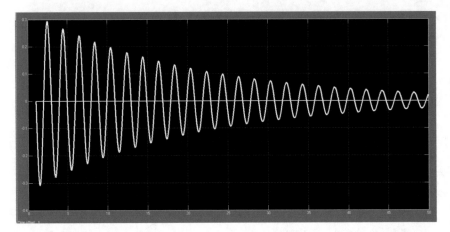

Fig. 11.5 Free object oscillations

Function convertor. A function convertor reflects the functional physic relationship (evaluated or obtained from experiments) of coordinate oscillations with periodic variations of the parameter. The relationship can be static and dynamic. A function generator is needed when, for example, either the dynamic parameter variation correction is required (e.g., by the introduction of derivative) or the parameter variation is performed by a dynamic unit (e.g., in a spring pendulum the pendulum length variation is performed using a mass–spring dynamic unit). In these cases, the function generator takes the form of the dynamic correction transfer function or dynamic unit.

Modulator. A modulator is realized by simple multiplication unit *Product*.

11.3 Parametric Regulator Structures

In this paragraph, we evaluate the performance of the suggested parametric regulators numerically on a second-order oscillatory object. Figure 11.5 depicts the free oscillations of the object without control. The envelope of the free oscillations tells that the uncontrolled object has the time constant $T = 25$ s.

Regulator 1. Figure 11.6a presents the model of oscillatory object which is controlled by an elementary parametric regulator. The amplifier and the delay unit τ are denoted as *Gain* and *Transport Delay*, correspondingly. The regulator output is connected to the coordinate control loop input through the *Product* modulator.

As follows from the oscillation plot in Fig. 11.6b, the regulator roughly triples the damping rate. The disadvantages are the existence of the constant component in parameter oscillation and the dependence of the parameter oscillation amplitude on the amplitude and frequency of the coordinate oscillations. This can result in the oscillation stability loss for certain parameters, as shown in Fig. 11.6c.

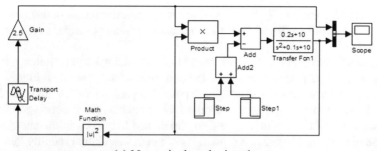

(a) Numerical analysis scheme

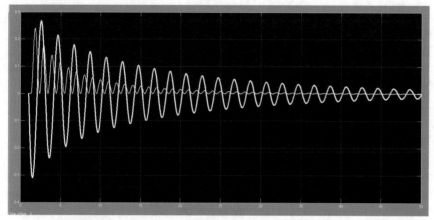

(b) Parametrically damped oscillation plot: $\Delta\ell=0.25$, $\tau=0.3$, Gain $= 2.5$, $T=10$ s

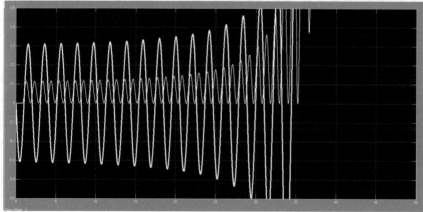

(c) Stability loss diagram: $\Delta\ell=0.15$, $\tau=0.8$ s, Gain $= 1.2$

Fig. 11.6 Parametric control simulation of regulator 1

Regulator 2. Figure 11.7 illustrates a different parametric regulator that contains the derivative du/dt. However, both the regulators are almost similar in respect of effectiveness.

Regulator 3. According to Figs. 11.7a and 11.8a, regulator 3 differs from regulator 2 by the order of frequency doubler and the derivative units in the control loop.

Figure 11.8 shows that the constant component of parameter oscillation has disappeared and the damping degree has raised. Nevertheless, the dependence of the parameter oscillation amplitude on the amplitude and frequency of the coordinate oscillations remains, which can potentially result in the oscillation stability loss.

Regulator 4. The oscillation amplitude boundary limiter (*Sign*) is added to the parametric control loop, as shown in Fig. 11.9.

Thus, the inserted limiter eliminates the parameter oscillation dependence on the amplitude and frequency of the output coordinate oscillations of the object. The transient period is four times shorter than that of free oscillations. It should be noted that the autoparametric control system is classified as the linear periodically nonstationary dynamic system, as follows from Fig. 11.9a.

11.4 Parametric and Coordinate Control

The section extends the results presented in [13] and discusses the scheme of combined parametric and coordinate control with internal (reactive) forces by elementary examples of the automatic autonomous object control.

In contrast to the coordinate control systems, those of parametric control are not much in use and less studied. In the parametrically controlled systems, the parameter, which value depends on the output object control coordinate, works as a regulator. The parametric control can be applied together with the coordinate control but there are autonomous devices such as pendulums, swings, cranes, separate traversers, and spacecraft orientation systems in which the parametric control is frequently the only possible approach.

Coordinate regulator. The reaction moment of a generic free inertia rotor motor is used, for example, to introduce internal control torque to an autonomous object. As soon as the rotor is switched on to rotate either clockwise or counterclockwise, an opposed fading reaction torque acts upon the stator fixed to the object. The known simple (neglecting the rotor inductance) mathematical model for a direct current (DC) motor is

$$J\frac{d\omega}{dt} = M = c_m i, \quad u = Ri + c_e \omega \tag{11.1}$$

and when the angular velocity is excluded, Eq. (11.1) delivers the transfer function

$$W_{en}(s) = \frac{M(s)}{u(s)} = \frac{k_{en}s}{T_{em}s + 1}, \tag{11.2}$$

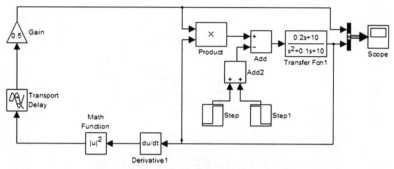

(a) Numerical analysis scheme

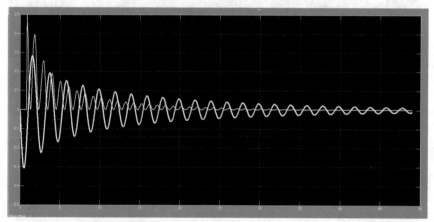

(b) Parametrically damped oscillation diagram: $\Delta\ell=0.5$, $\tau=0.8$ s, $T=10$ s

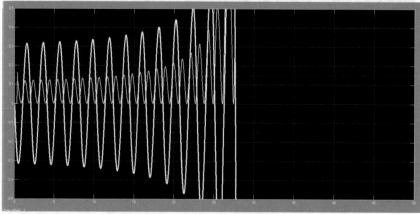

(c) Stability loss diagram: $\Delta\ell=0.13$, $\tau=0.3$ s.

Fig. 11.7   Parametric control simulation of regulator 2

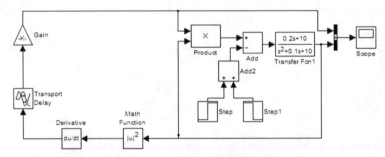

(a) Numerical analysis scheme

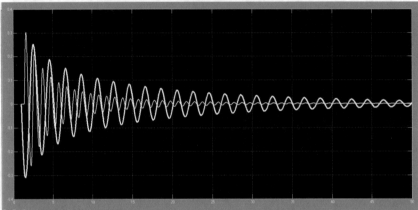

(b) Parametrically damped oscillation diagram: $\Delta\ell$=0.3, τ=0.4 s, *Gain* 1.0, T=8 s

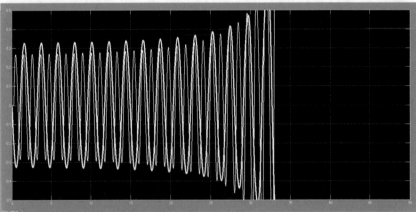

(c) Stability loss diagram: $\Delta\ell$=0.25, τ=0.23 s, *Gain* 0.88

Fig. 11.8 Parametric control simulation of regulator 3

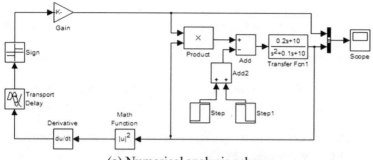

(a) Numerical analysis scheme

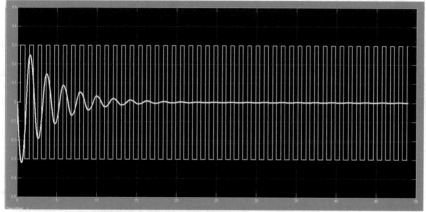

(b) Parametrically damped oscillation diagram: $\Delta\ell$=0.3, τ=0.4 s, *Gain* 0.3, T=6 s

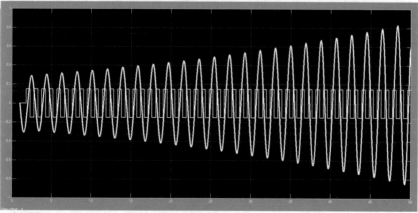

(c) First parametric resonance oscillation diagram: $\Delta\ell$=0.15, τ=0.8, *Gain* 0.15

Fig. 11.9 Parametric control simulation of regulator 4

where J is the rotor inertia moment; M is the stator reaction torque; u and I are voltage and current of the motor, correspondingly; c_m and c_e are the electromagnet and electric constants of the motor, correspondingly; $k_{en} = j/c_e$ is the transfer constant of the stator reaction moment; and T_{em} is an electromechanical time constant of the motor.

As one can easily observe, the motor works as a differentiator in this application; it can modify the control function correction together with the control torque generation.

Let us examine numerically the quality of the combined parametric and coordinate control of simple autonomous objects.

Stable pendulum. In the elementary example, a stable autonomous mathematical pendulum is assumed to have been described by the classical model.

$$ml^2\ddot{\alpha} + \zeta\dot{\alpha} + mgl\alpha = M, \tag{11.3}$$

where M is a certain initial torque. When the mass $m = 1$ kg, the length $l = 1$ m, the damping factor is $\zeta = 0.05$ Ns/rad, and the gravitational acceleration $g \cong 10$ m/c^2, the pendulum transfer function looks like

$$W(s) = \frac{\alpha(s)}{M(s)} = \frac{1}{s^2 + 0.1s + 10}.$$

The time constant T of the damped oscillation envelop of a free pendulum with the above specified parameters is 25 s. Let us compare this value to what we can reach with the parametric and coordinate control of the pendulum.

Figure 11.10 plots the parametric pendulum control model based on the model of parametric regulator 4. In this case, the pendulum model under the same parameter values takes the form

$$m[l + \Delta l(t)]^2\ddot{\alpha} + \zeta\dot{\alpha} + mg[l + \Delta l(t)] = 0. \tag{11.4}$$

Having assumed small oscillations $\Delta l(t) \ll 1$, we linearize the equation and derive the transfer function of a parametrically controlled pendulum as

$$W(s) = \frac{\alpha(s)}{\Delta l(s)} = \frac{0.2s + 10}{s^2 + 0.1s + 10}. \tag{11.5}$$

As appears from Fig. 11.10, the time constant T of the free oscillation envelope is about 7 s now.

Next, let us consider the pendulum coordinate control with the help of the reactive moment of the free rotor motor fixed to the pendulum. Figure 11.11 shows the relevant block scheme.

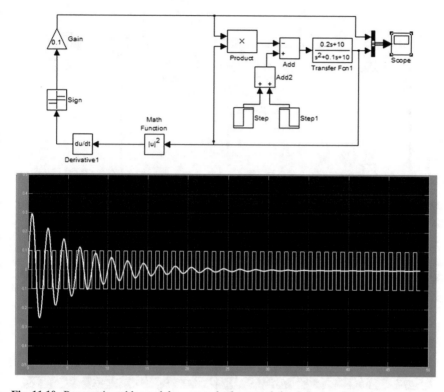

Fig. 11.10 Parametric stable pendulum control scheme and simulation results

The controlling torques of parametric and coordinate control are chosen to be about equal. As appears from Figs. 11.10 and 11.11, those free oscillation processes are nearly the same.

Finally, Fig. 11.12 reveals the model of combined coordinate-parametric control for a gravitation spring pendulum. The denominator of the transfer function *Transfer Fcn2* is $0.2s + 10$. The initial conditions are set by an impulse coming from the sum unit *Add2* at the time $t = 1$ s with duration 0.3 s. The simulation shows that the system has a minimal time constant $T = 4$ s.

Unstable pendulum. Let us consider the parametric stabilization of the unstable Kapitza pendulum. The unstable equilibrium that is the inverted pendulum with the joint below the center of gravity is stabilized by the vibration of the joint or support. Let us agree on the mathematical model of the process in the form

$$\ddot{\alpha} - [(g + a\Omega^2 \sin \Omega t)/\ell]\alpha = 0,$$

where the variable component of the parameter or acceleration g is harmonically varied with frequency Ω and amplitude $a\Omega^2$.

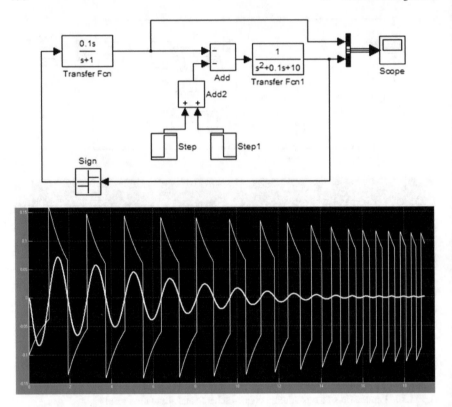

Fig. 11.11 Model and diagram of free pendulum oscillations controlled by reactive motor torque

The first parametric resonance excitation condition at the frequency $\omega = \Omega/2$ and the first condition for inverted pendulum stabilization

$$a\Omega < \frac{l\Omega}{2} + \frac{2g}{\Omega} \tag{11.6}$$

were obtained in Sect. 5.1.

The second condition of the inverted pendulum stabilization requires the vibration speed to be greater than the gained speed of mass at its free fall from the height of the pendulum length, i.e., $a\Omega > \sqrt{2gl}$. The required high speed and acceleration vibration lead to the high excitement threshold and modulation depth (of acceleration).

In this sense, the problem of inverted pendulum stabilization is the only case in the book that requires the justification of the applied approximate models and methods.

Figure 11.13 presents the block scheme for the simulation and some results.

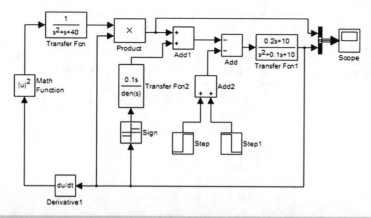

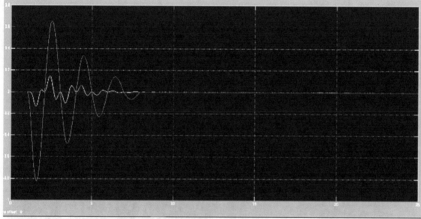

Fig. 11.12 The block scheme for the numerical analysis and the results of simulation for the swing pendulum combined parametric-coordinate control

It should be noted that the parametric stabilization of an inverted pendulum by means of the foundation vibrations is achieved, in principle, by an external action. Let us consider the autonomous stabilization system of an inverted pendulum with a reactive motor torque.

Figure 11.14 displays the model of a linear automatic stabilization system with an unstable pendulum controlled by the reactive stator moment, while the motor rotor is free.

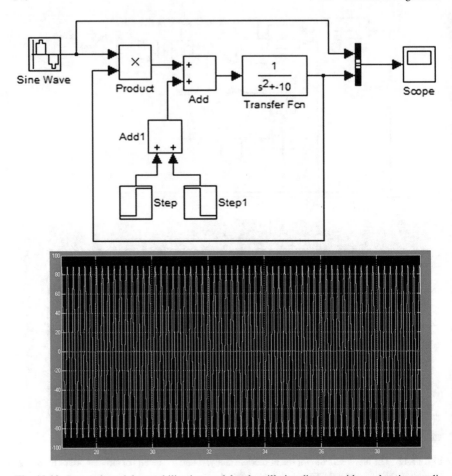

Fig. 11.13 Inverted pendulum stabilization model and oscillation diagram with acceleration amplitude of 60 m/c^2 and frequency of 11 rad/s

The stabilization is achieved due to the presence of an unstable feedback correction unit with transfer function *Transfer F2* and the motor stator torque with the transfer function *Transfer F*. The *Gain* coefficient is set to be 100. An initial impulse is formed by the units *Step* and *Step 1*.

We can also prove the concept of a simplified nonlinear oscillatory regulator for an unstable platform with a motor. Figure 11.15 depicts the platform model.

The initial deviation is ensured by the *Step* unit (step time of 0.1, an impulse value of 0.1) and the *Step 1* unit (step time of 0.2, an impulse value of 0.1). A real motor delay is assumed to be of 0.01 s. The system falls into self-oscillating motion mode.

In general, the quality of transient responses of the coordinate and parametric control to the autonomous objects seems to be the same.

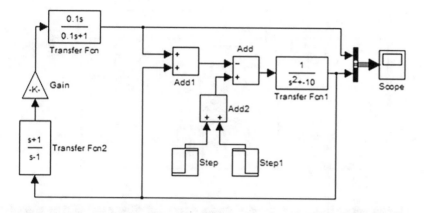

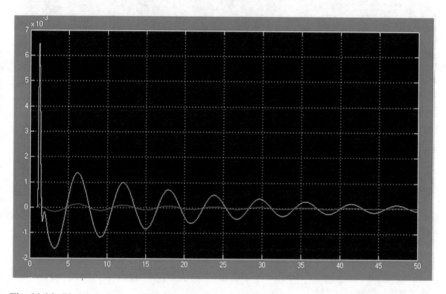

Fig. 11.14 Linear system model of automatic coordinate stabilization for inverted pendulum and its transient response

The inverted Kapitsa's pendulum [10] is an example of structurally unstable and non-oscillatory plant. The controllers of Figs. 11.14 and 11.15 transform it into a stable oscillatory system. Let us study a general case of parametric stabilization of unstable object. We use an automatic tracking system drive as an example of parametric stabilization. The drive is assumed to have the transfer function

$$W_{ol}(s) = \frac{1}{(s^2 + s + 9)s}.$$

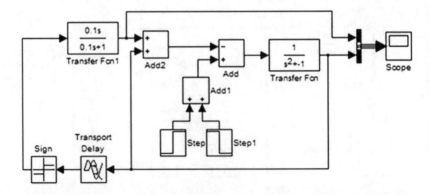

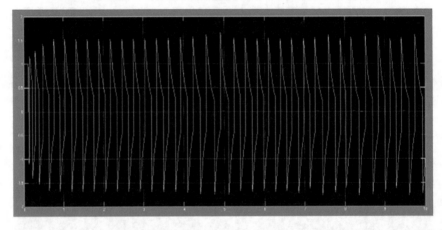

Fig. 11.15 Coordinate regulator operation diagram and coordinate nonlinear automatic system model for unstable pendulum stabilization

A negative non-inertial feedback with the gain k loops the drive. Now, the closed-loop transfer function is

$$W_{cl}(s) = \frac{1}{s^3 + s^2 + 9s + k}.$$

The system's natural frequency is $\omega_o = 3$ rad/s. At the parameter values $k > k_{cr} = 9$, the tracking system is unstable. Let us take $k = 9.5$ and introduce the parametric controller. The scheme of the final closed-loop system is given in Fig. 11.16.

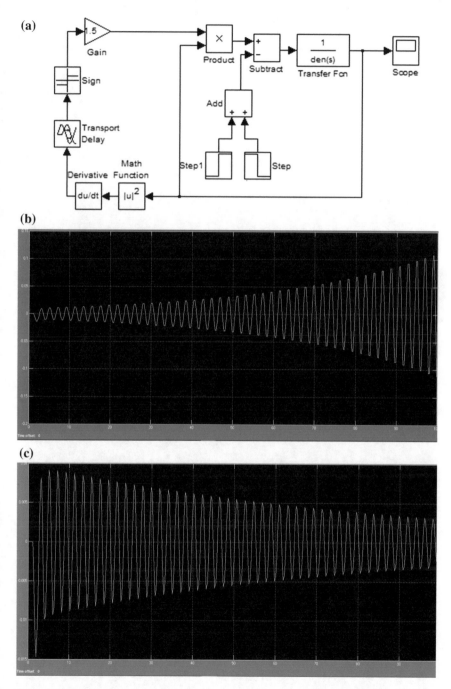

Fig. 11.16 The block diagram of the tracking system. **a** The transient response of the system without parametric control. **b** The transient with the parametric regulator. **c** $\tau = 0.3$ s, $\Delta \ell = 1.5$

Part V
Applications

The part provides examples of parametric excitation condition analysis in systems of various fields: mechanical, hydraulic, aerodynamic, electric, radio, and even economic.

Chapter 12
Mechanical Problems

12.1 Bending Oscillations of Elastic Beam

The vertical oscillations of a horizontal dead-end elastic beam of rectangular but substantially non-square cross-section are considered. Assuming the beam cross-section width is much greater than its thickness, horizontal stiffness of the beam is taken to be infinitely large.

The approximate description of the bent oscillations of a horizontal spring beam can be represented by a discrete model in the form of a n-segment lumped circuits

$$
\begin{aligned}
Q(n-1) - Q(n) &= -ms^2 y(n) \\
M(n-1) + Q(n-1)\ell &= M(n) \\
M(n-1) &= -c[\theta(n-1) - \theta(n)] \\
\Theta(n) &= [y(n) - y(n-1)]/\ell,
\end{aligned}
\tag{12.1}
$$

here, Q, M, θ, and y are the external vertical force, external moment, deflection angle, and vertical deflection of the segment, correspondingly; ℓ, c, and m are the distributed parameters: length, stiffness, and mass, correspondingly. The discrete system equations are obtained using z-transform as

$$
\begin{aligned}
(z^{-1} - 1)Q(z) + ms^2 Y(z) &= -Q(0) \\
z^{-1}\ell Q(z) + (z^{-1} - 1)M(z) &= -M(0) \\
\theta(z)(z^{-1} - 1) + c^{-1}z^{-1}M(z) &= -\theta(0) \\
\ell\theta(z) - (1 - z^{-1})Y(z) &= -Y(0).
\end{aligned}
\tag{12.2}
$$

The boundary conditions are on the right side of equations (12.2). The system determinant has the form

© Springer International Publishing AG, part of Springer Nature 2017
L. Chechurin and S. Chechurin, *Physical Fundamentals of Oscillations*,
https://doi.org/10.1007/978-3-319-75154-2_12

$$\Delta(z) = -\frac{c}{z^4}\Delta_1\Delta_2$$

$$= \frac{c}{z^4}\left[z^2 - 2z\left(1 + \frac{\ell s}{2}\sqrt{\frac{m}{c}}\right)\right]\left[z^2 - 2z\left(1 - \frac{\ell s}{2}\sqrt{\frac{m}{c}}\right)\right] \quad (12.3)$$

System (12.2) takes the matrix notation

$$X(z) = H(z, s)X(0), \quad X = (Q, M, \theta, Y)^T \quad (12.4)$$

Assuming zero boundary conditions $M(0) = \theta(0) = Y(0) = 0$, the output coordinate solution $Y(z, s)$ is written according to Kramer's rule as

$$Y(z, s) = \frac{l^2 z^{-2}}{\Delta(z, s)}Q(0) = -\frac{l^2 z^2}{c\Delta_1\Delta_2}Q(0) \quad (12.5)$$

To return from the z-transformation forms to the originals, expression (12.5) is decomposed into

$$Y(z, s) = -\frac{l^2 z^2}{c\Delta_1\Delta_2} = -\frac{l^2}{2c}\frac{1}{(ch\beta_1 - ch\beta_2)}\left(\frac{z}{\Delta_1} - \frac{z}{\Delta_2}\right), \quad (12.6)$$

where we denote $ch\beta_1 = 1 + \frac{\ell s}{2}\sqrt{\frac{m}{c}}$ and $ch\beta_2 = 1 + \frac{\ell s}{2}\sqrt{\frac{m}{c}}$.
Using the inverse z-transformation (see Table 1.1),

$$Y(n, s) = \left[\frac{l^2}{2c(ch\beta_2 - ch\beta_1)}\left(\frac{sh\beta_1 n}{sh\beta_1} - \frac{sh\beta_2 n}{sh\beta_2}\right)\right]Q(0) \quad (12.7)$$

Function (12.7) determines the forced bending oscillations of the coordinate Y as a result of excitation of the beam by the external force Q. If the beam rotates at a fixed rate, the external force is absent, but the bending stiffness c varies periodically. The free oscillation equation is derived by setting to zero the expression in the square brackets of (12.7)

$$\frac{1}{(ch\beta_2 - ch\beta_1)}\left(\frac{sh\beta_1 n}{sh\beta_1} - \frac{sh\beta_2 n}{sh\beta_2}\right) = 0 \quad (12.8)$$

in which the stiffness parameter varies harmonically

$$c(t) = c + \tilde{c}(t) = c + \Delta c \sin 2\omega t, \quad (12.9)$$

where $\tilde{c}(t) \ll c$ and ω is the shaft rotation rate.

The bigger the number of segments the better the accuracy of (12.9). Assuming, for example, $n = 6$, after some trigonometry, we can arrive at the following free oscillation equation in operator form

$$\ell^4 m^2 s^4 + 36\ell^2 mc(t)s^2 + 35c^2(t) = 0 \tag{12.10}$$

Taking into account the stiffness periodic variations, Eq. (12.10) takes the form

$$\ell^4 m^2 s^4 + 36\ell^2 mcs^2 + 35c^2 \cong -(36\ell^2 ms^2 + 70c)\Delta c(t) \tag{12.11}$$

According to the harmonic stationarization method, the variable parameter $\tilde{c}(t)$ has the complex gain $0.5\,|\Delta c|\exp j\varphi$ at the coordinate oscillations at the first parametric resonance frequency ω. The latter means that the resonance excitation condition in the amplitude–phase–frequency characteristic plane appears as

$$W_b(s)_{s=j\omega} = \frac{36\ell^2 ms^2 + 70c}{\ell^4 m^2 s^4 + 36\ell^2 mcs^2 + 35c^2} = 2\,|\Delta c|^{-1}\,e^{j\varphi} \tag{12.12}$$

Let us analyze the result numerically assuming that the following beam parameters $m_0 = 1$ kg, $l = 0.4$ m, $c = 40$ Nm/rad are distributed among six segments. Thus, the amplitude balance equation is

$$W_b(s)_{s=j\omega} = \frac{6s^2 + 4800}{0.026s^4 + 230s^2 + 56{,}000} = 2\,|\Delta c|^{-1}\,e^{j\varphi} \tag{12.13}$$

The left side of (12.13), or amplitude–phase–frequency characteristic, has its modulus equal of 20 at the first resonance frequency of 15.6 rad/s. As follows from the modulus equality of condition (12.13), the threshold modulus of the stiffness oscillations is about $|\Delta c| \cong 0.1$ Nm/rad on the frequency of 32.2 rad/s. The result is confirmed by the numerical simulation given in Fig. 12.1.

12.2 Torsional Oscillations

Let us consider the torsional oscillations of an elastic beam of rectangular profile with one free one fixed end. The finite-differential equations for the n-th segment have the elementary form

$$\begin{aligned}
J\ddot{\varphi}(n) &= M(n) - M(n+1) \\
M(n+1) &= c[\varphi(n) - \varphi(n+1)]
\end{aligned} \tag{12.14}$$

The coordinate images are obtained using a z-transformation as

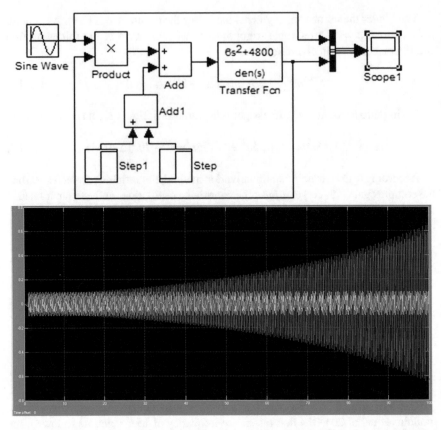

Fig. 12.1 Model simulation set up and parametric excitation diagram of bending oscillations by elastic rectangular beam rotation; the initial start pulse duration is $\tau = 0.1$ s; stiffness oscillation amplitude $\Delta c = 0.1$ Nm/rad

$$M(z) = \frac{c(1-z)}{z}\varphi(z) + M(0) + c\varphi(0)$$

$$\varphi(z) = \frac{1 - zM(z) + zM(0)}{Jp^2}$$

Excluding $\varphi(z)$ from the above coordinate images, subject to the boundary condition $\varphi(0) = 0$ the relation is transformed to

$$M(z) = \frac{z^2 - z(2ch\xi - 1)}{z^2 - 2zch\xi + 1}M(0).$$

The approximate solution is found by applying the inverse z-transformation

$$M(p, n) = \left[ch\xi n - (ch\xi - 1)\frac{sh\xi n}{sh\xi} \right] M(0), \qquad (12.15)$$

where $ch\xi = 1 + Jp^2/2c$.

Some manipulations can yield the transfer function for the variable parameter $c(t) = c + \Delta c \sin \Omega t$ with $n = 4$ as an example

$$W(s) = \frac{5J^2s^4 + 12Jcs^2 + 3c^2}{J^3s^6 + 5J^2cs^4 + 6Jc^2s^2 + c^3}. \qquad (12.16)$$

The distributed inertia moment and torsional stiffness are taken to be $J = 0.005$ kg m^2, $c = 20$ Nm/rad. Having omitted small terms and introduced a small damping as a term at operator s in the denominator, the final condition of the first parametric resonance excitation appears as

$$
\begin{aligned}
W(s)_{s=j\omega} &\approx \frac{12Jcs^2 + 3c^2}{5J^2cs^4 + 6Jc^2s^2 + s + c^3} \\
&\cong \frac{1.2s^2 + 1200}{0.0025s^4 + 12.5s^2 + s + 8000} \\
&= 2\,|\Delta c|^{-1}\, e^{-j\varphi}
\end{aligned}
\qquad (12.17)
$$

Frequency characteristic (12.17) has the modulus $A = 20$ at the first resonant frequency $\omega = 25$ rad/s. The first parametric resonance excitation is observed at the parameter oscillation amplitude $\Delta c > 0.1$ Nm/rad and the frequency $\Omega = 50$ rad/s, as shown in Fig. 12.2.

12.3 Bending-Torsional Flatter

In most of analyzed situations of parametric resonance excitation, the parameter variation profile was given as an external signal so far. An exception are several problems of Part 4. Strictly speaking, the case of parametric oscillations like bending or torsional ones cannot be called self-excitation. At the same time, coexistence of bending and torsional oscillations can lead to and often results in a self-excitation or so-called flatter. A recent example of a parametric self-excitation was recently observed in the form of bending-torsional oscillations of the seven-kilometer bridge, which was built across the Volga River near Volgograd (former Stalingrad) in 2010 (Fig. 12.3).

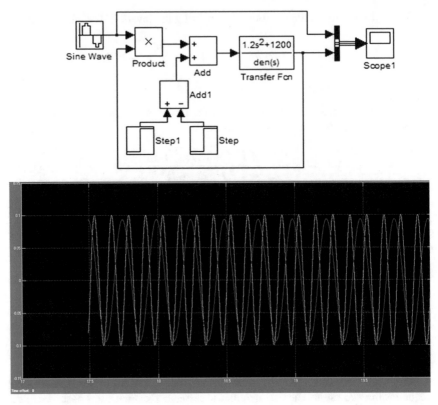

Fig. 12.2 Model for simulation and excitation process of torsional parametric oscillations; $\Delta c = 0.1$ Nm/rad, $\Omega = 50$ rad/s, initial unit pulse duration is 0.1 s

Let us consider one of the simplest cases of flatter in the form of two-circuit parametric self-excitation or bending-torsional flatter; the model is shown in Fig. 12.4.

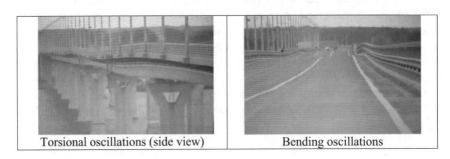

Torsional oscillations (side view)   Bending oscillations

Fig. 12.3 Volgograd bridge oscillations

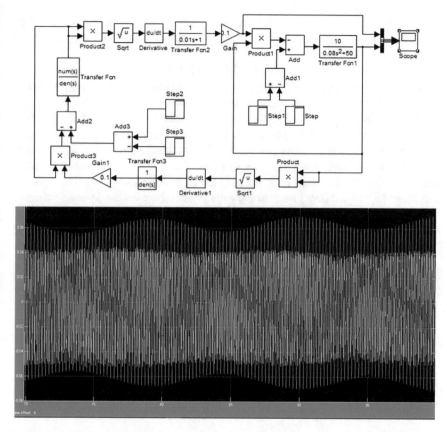

Fig. 12.4 Block diagram of two-circuit parametric flatter and self-excitation plot. Start pulse duration is $\tau = 1$ s with unit amplitude 1 of *Step* or *Step1* and with amplitude 2 of *Step2* or *Step3*

In the block diagram model, the torsional oscillation loop includes *Transfer Fcn* function (12.17) and the bending oscillation loop has the *Transfer F1* function derived based on description (12.8) concerning the four-segment model with distributed parameters $c = 5$ Nm/rad, $\ell = 0.4$ m, $m = 0.5$ kg in the form

$$W_b(s)_{s=j\omega} \cong \frac{10}{\ell m s^2 + 10c} = \frac{10}{0.08s^2 + 50}. \tag{12.18}$$

The resonance frequency and modulus of the frequency response are roughly the same to those of torsional oscillations (12.17).

The assumed transfer functions to torsional (12.17) and bending (12.18) oscillations in respect to the same beam are parametrically connected by their stiffnesses through the frequency doublers (i.e., the *du/dt* units, the *Product* and *Product 2* multiplier units, the *Sqrt* and *Sqrt1* square root units, the *Transfer2*, and *Transfer3* filters) and the *Gain-0.1* and *Gain1-0.1* coupling coefficient units. The high-frequency filter has no influence on excitation processes and just shapes the oscillation form. The *Product 1* and *Product 3* multipliers make the product of stiffness and coordinate oscillations.

The self-excitation process presented in Fig. 12.4 lasts for several hundred seconds under the specified minor coupling coefficients, the instability, and asymptotically unlimited growth of the coordinate occurs after that.

The time interval from the beginning of excitation to the parametric excitation process shutdown depends on initial (external) conditions and the excess of the excitation threshold by the coupling coefficients. The excitation thresholds are overcome by the external disturbances. They are simulated by means of short pulses from the *Step2* and *Step3* units. Last, in addition to the initial conditions, the time moment corresponding to stability loss also depends substantially on the sampling time of numerical simulation. The main cause of the stability loss is that the amplitude of parameter oscillations depend linearly on the coordinate oscillation amplitude; the system becomes nonlinear as the oscillation amplitudes grow.

Chapter 13
Parametric Resonance in Fluid Dynamics

In numerous complex problems of fluid dynamics, either the motion of a medium (liquid, gas, and air) in which an object is situated or the motion of an object (aircraft, a rocket, and a ship) inside the medium as well as medium-object interference are considered. As a rule, those problems cannot be analytically solved. They are addressed numerically and by physical modeling of the medium and the object.

13.1 Parametric Resonance in Hydrodynamics

Let us consider a simple example. We study "strange" periodically repeated oscillations of a river buoy in a strong flow. Such oscillations are observed when the submerged buoy sometimes is under sometimes on the surface. It turned out that the forced oscillations of the submerged buoy initiated, for example, under a stall flutter is able to cause a parametric resonance which, in turn, gives a rise to the amplitude jump of initial forced oscillations. The initiation of the hidden parametric oscillations deals with the fact that the torque M, pushing the buoy out of the water, depends nonlinearly on the deviation angle as follows

$$M = M_0\alpha + M_1\alpha^3.$$

So, with forced angle oscillations $\alpha(t) = A \sin \omega t$, the parameter or torque rate varies periodically as

$$dM/d\alpha|_{\alpha(t)} = M_0 + 3M_1 A^2 \sin^2 \omega t = M_0 + 1.5M_1 A^2(1 + \cos 2\omega t).$$

It seems that the first, "hidden", parametric resonance at the forced oscillation frequency ω can arise at the double parameter variation frequency.

The equation of an ordinary pendulum with damping ξ is used to a waterproof cylinder, which is tied to a rope and submerged (for example, due to a spring water flood)

© Springer International Publishing AG, part of Springer Nature 2017
L. Chechurin and S. Chechurin, *Physical Fundamentals of Oscillations*,
https://doi.org/10.1007/978-3-319-75154-2_13

$$Js^2\alpha + \xi s\alpha + [M_0 + \Delta M(\alpha)] = 0,$$

here, M_0 is the initial buoyancy torque at $\alpha = 0$, $\Delta M(\alpha)$ is the buoyancy torque increment connected with an immersion at the deviation, α. By unfolding the torques, the last equation can be written in the form

$$Js^2\alpha + \xi s\alpha + [(F_0 - P)\ell + \Delta F(\alpha)\ell] \sin \alpha = 0,$$

where F_0 is the initial buoyancy force at $\alpha = 0$, $\Delta F(\alpha)$ is the buoyancy force increment owing to the oscillations of the cylindrical buoy submerged ℓ is the rope length, P is the buoy weight.

Let us introduce the following parameters concerning the cylinder: the base area $q(a)$, the initial immersion h_0, the specific liquid weight d, and the deviation immersion $\Delta h(\alpha) = \ell(1 - \cos \alpha)$. Then,

$$Js^2\alpha + \xi s\alpha + \ell[(qh_0d - mg)\sin \alpha + qd\ell(1 - \cos \alpha)\sin \alpha] = 0.$$

In the notations, $c_0 = \ell J^{-1}(qh_0d - mg)$, $c_1 = J^{-1}qd\ell^2$ the equation takes the form

$$s^2\alpha + bs\alpha + c_0 \sin \alpha + c_1(\sin \alpha - 0.5 \sin 2\alpha) = 0.$$

Linearizing the last nonlinear equation up to α^3, the approximate equation for the small deviations $|\alpha > |\alpha^3||$ is obtained in the form

$$s^2\alpha + bs\alpha + c_0\alpha + c_1\frac{\alpha^3}{2} = 0.$$

The approximate equation is known as Duffing equation. The tabular format of the harmonic linearization coefficient for a nonlinear function is as

$$W_h(A) = 3c_1A^2/8.$$

Then, the radius and shift of the parametric resonance circle center are as follows:

$$\rho_1 = |a_1(A)| = \frac{A}{2}\left|\frac{dW(A)}{dA}\right| = 3c_1A^2/8,$$

$$a_0(A) = W(A) + \frac{A}{2}\frac{dW(A)}{dA} = 3c_1A^2/4.$$

The inversed amplitude–phase–frequency characteristic of the linear part

$$W^{-1}(j\omega) = (j\omega)^2 + b(j\omega) + c_0 = -\omega^2 + jb\omega + c_0$$

Fig. 13.1 Parametric resonance excitation condition

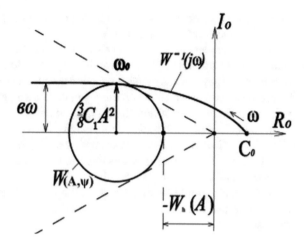

together with the parametric resonance circle are plotted in the plane (R_0, I_0), as shown in Fig. 13.1.

It is clear from Fig. 13.1 that the frequency ω_0, and the amplitude, A_0, from which the parametric resonance excitation begins can be derived from the equalities

$$2\mathrm{Re}W^{-1}(j\omega) = \mathrm{Im}W^{-1}(j\omega) = a_1(A),$$

and the results can be given the form

$$\omega_0 = b + \sqrt{b^2 + c_0}, \quad A_0^2 = \frac{8}{3}\frac{\omega_0 b}{c_1}.$$

Hence, the excitation conditions in terms of the buoy parameters, that is to say, the conditions of parametric resonance excitation and the forced oscillation stability loss can be expressed as

$$A^2 \geq \frac{8}{3}\frac{b\omega_0}{c_1}, \quad \omega > \omega_0 \cong b + \sqrt{b^2 + c_0}.$$

From the formal mathematics standpoint, there happens to be an amplitude jump to infinity of the forced oscillation amplitude. Of course as a physical matter, the infinite jump cannot be realized. The fact is that the assumed mathematical buoy formulation ceases to be true as soon as the buoy lays on the water surface. During the parametric oscillations, the buoy tilts on the water surface under the action of the flow before its submersion comes to the end. Once the buoy spends its motion energy completely, it extrudes out of the water again. When the water flow is not strong enough, i.e., with still water, the buoy parametric resonance is also feasible in any vertical plane passing through the suspension axis. But even there, the forced oscillation jumps as large as an infinity value cannot be realized since the "parametric" force vanishes as soon as the vertical buoy submerges totally and the parametric resonance circle radius becomes zero by virtue of $\Delta h = 0$.

A careful reader can immediately ask a question "And where are the forces which cause forced oscillations?" There can be several reasons for occurrence the forced oscillations. They are, e.g., wind gusts, wind waves, and wave formation from moving ships. Moreover, at watercourse, there can be known periodic separations of water vortexes in the process of flow the submerged buoy.

In conclusion, assuming the rough buoy parameters $P = 50$ kg, $F_0 = 55$ kg, $l = 10$ m, $J = 500$ kg m s^2, $s = 0.3$ m^2, $b = 0.1$ s^{-1}, $c_0 = 0.05$ s^{-2}, $c_1 = 0.06$ s^{-2}, the following results close to the reasonable ones can be obtained as $\omega_0 \approx 0.35$ rad s^{-1} and $A_0 \approx 0.5$ rad.

In whole, it should be noted that oscillation processes in liquids have a complex nature and most often combine forced parametric oscillations and nonlinear self-oscillations.

13.2   Parametric Resonance in Aerodynamics

As a rule, an aircraft wing has modest bending stiffness and is able to make minor oscillations in the vertical plane. Sometimes, the oscillations can be watched in the side aircraft window. The torsional wing stiffness along the longitudinal axis is much greater, so special devices can only sense the torsional oscillations.

An aircraft wing model can be simplified to a beam one free on clamped end. The beam stiffness is provided to correspond to the wing-averaged torsional or bending stiffness.

Again, assuming the similarity with a spring pendulum, one can conclude that flexure-torsional flutter is feasible: under certain conditions, the forced bending oscillations can initiate the parametric torsional oscillations.

The similarity between wing oscillations and the bending-torsional oscillations of a spring pendulum is far from a complete one. The fact is that with high speeds in respect to the surrounding air or aqueous media, even small torsional oscillations of wings result in the sharp fluctuations of a lifting force. That is why, the parametric resonance is excited much easily and may significantly endanger the ride. In other words, in that case the parametric resonance has a coordinate-parametric origination. Hence, the model presented in Fig. 13.2 can be built by adding a coordinate connection from the torsional coordinates to the bending ones through the *Gain2* amplifier with a gain of 0.5 to the model in Fig. 12.4. The self-excitation diagrams are also given in Fig. 13.2 to be compared with the bended-torsional flatter shown in Fig. 13.1. The stability loss is observed after 40 s from the beginning of the self-excitation process.

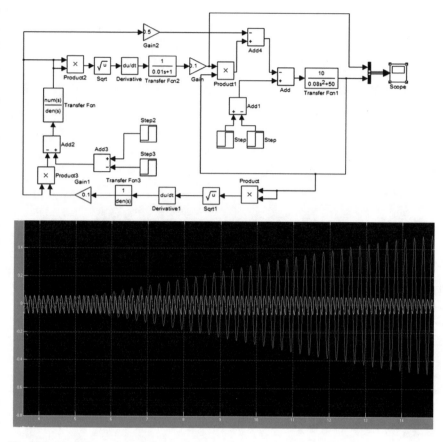

Fig. 13.2   Block diagram simulation model and self-excitation plot for coordinate-parametric-and-bending-torsional flatter in moving medium

Chapter 14
Electro and Radio Engineering Problems

14.1 Variable Parameter Circuits

The excitation of parametric resonance oscillations are known to be feasible in linear electric circuits with periodically variable parameters. This phenomenon is widely used in radio engineering for generators and parametric amplifiers design.

Let us consider the elementary electric circuit presented in Fig. 14.1. The circuit consists of the following time-varying parameters: inductance $L(t)$, capacitance $C(t)$, and resistance $R(t)$.

When a voltage source is absent, the sum of voltages at the circuit elements is zero

$$u_L(t) + u_R(t) + u_C(t) = 0.$$

All the voltages depend on the circuit current as follows

$$\frac{d[L(t)i(t)]}{dt} + R(t)i(t) + \frac{1}{C(t)} \int i(t)dt = 0 \qquad (14.1)$$

Let us deduce the excitation conditions in the circuit shown in Fig. 14.1 for the cases when the inductance or th capacitance or resistance vary.

14.1.1 RLC Circuit with Periodically Varying Inductance

Equation (14.1) takes the form

$$\frac{d[L(t)i(t)]}{dt} + Ri(t) + \frac{1}{C} \int i(t)dt = 0$$

and after differentiation, it becomes as

© Springer International Publishing AG, part of Springer Nature 2017
L. Chechurin and S. Chechurin, *Physical Fundamentals of Oscillations*,
https://doi.org/10.1007/978-3-319-75154-2_14

Fig. 14.1 Variable
parameter electric circuit

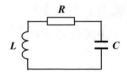

$$\frac{d^2[L(t)i(t)]}{dt^2} + R\frac{di(t)}{dt} + \frac{1}{C}i(t) = 0.$$

Transferring to the operator form

$$s^2[L(t)i(t)] + Rs\,[i(t)] + \frac{1}{C}i(t) = 0.$$

Let the inductance L vary according to the profile

$$L(t) = L_0(1 + a\,\cos\Omega t),\, 0 \le a \le 1.$$

Let us move from the operator equation to the frequency one by substituting s by $j\Omega/2$ and the periodic inductance $L(t)$ by the transfer function $W(j\varphi)$:

$$\left(\frac{j\Omega}{2}\right)^2 W(j\varphi) + 2\alpha\left(\frac{j\Omega}{2}\right) + \omega_0^2 = 0, \tag{14.2}$$

where $\alpha = \frac{R}{2L_0}$, $\omega_0^2 = \frac{1}{L_0C}$.

According to the given parameter variation law, the linearized frequency equation is found as a result of the substitution of the parameter transfer function $W(j\varphi)$ in the form of (4.13) into the frequency equation.

$$\left(\frac{j\Omega}{2}\right)^2 \left(1 - \frac{a}{2}e^{-j\varphi}\right) + 2\alpha\left(\frac{j\Omega}{2}\right) + \omega_0^2 = 0.$$

Excluding φ, the condition of parametric resonance excitation in the circuit is found as

$$a = \frac{8}{\Omega^2}\sqrt{\left(\omega_0^2 - \frac{\Omega^2}{4}\right)^2 + \alpha^2\Omega^2}.$$

The geometrical approach to the solution is easy too. The inverse frequency response comes out of equation (14.2) as

$$W^{-1}(j\omega) = \frac{2\alpha(j\omega) + \omega_0^2}{(j\omega)^2}.$$

Fig. 14.2 On evaluation of
first parametric resonance
conditions: variable
inductance case

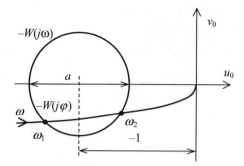

Figure 14.2 depicts the rough frequency characteristic view and the parametric
resonance circle.

As appears from Fig. 14.2, the excitation occurs in the frequency range $2\omega_1 \leq \Omega \leq 2\omega_2$. It is important to note that the frequency response can be obtained from
not only mathematical modeling but from the physical experiment too. A harmonic
generator replaces the inductance in the electric circuit for this. The amplitude and
phase of the current are measured at every frequency. The measured experimental
frequency response $W_{cl}(j\omega)$ has to be multiplied by $j\omega L_0$ because

$$W(j\omega) = j\omega L_0 W_{cl}(j\omega).$$

14.1.2 RLC Circuit with Periodically Varying Capacitance

The following capacitance variation profile is assumed

$$\frac{1}{C} = \frac{1}{C_0}(1 + a \cos \Omega t), \quad 0 \leq a \leq 1.$$

In this case, Eq. (14.1) takes the form

$$L\frac{d[i(t)]}{dt} + Ri(t) + \frac{1}{C(t)}\int i(t)dt = 0.$$

Replacing the variable of current $i(t)$ by the charge $q(t) = \int i(t)dt$ and denoting

$$\alpha = \frac{R}{2L}, \omega_0^2 = \frac{1}{LC_0},$$

the initial equation

$$\frac{d^2[q(t)]}{dt^2} + 2\alpha\frac{dq(t)}{dt} + \omega_0^2(1 + a \cos \Omega t)q(t) = 0$$

Fig. 14.3 On evaluation of
the first parametric
resonance conditions.
Variable capacitance case

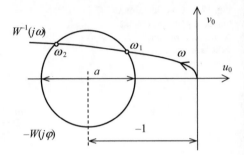

is given the operator form

$$s^2[q(t)] + 2\alpha s q(t) + \omega_0^2 (1 + a \, \cos \Omega t) q(t) = 0.$$

Replacing the periodic parameter by its transfer function (4.13) with a 90° turn
in respect of a cosine function, the frequency equation with $s = j\Omega/2$ becomes

$$\left(\frac{j\Omega}{2}\right)^2 + 2\alpha \left(\frac{j\Omega}{2}\right) + \omega_0^2 \left(1 - \frac{a}{2} e^{-j\varphi}\right) = 0.$$

Hence the parametric oscillation excitation condition is

$$a = \frac{2}{\omega_0^2} \sqrt{\alpha^2 \Omega^2 + \left(1 - \frac{\Omega^2}{4}\right)^2}.$$

The inverse frequency response of the stationary part of the circuit is

$$W^{-1}(j\omega) = \frac{(j\omega)^2 + 2\alpha(j\omega)}{\omega_0^2}.$$

The graphical solution in the plane of the inverse frequency response is shown in
Fig. 14.3.

14.1.3 RLC Circuit with Periodically Varying Resistance

In this case, Eq. (14.1) takes the form

$$L \frac{d^2[i(t)]}{dt^2} + \frac{d\,[R(t)i(t)]}{dt} + \frac{1}{C} i(t) = 0$$

And the frequency response of the stationary part of the circuit is

Fig. 14.4 On evaluation of
the first parametric
resonance conditions.
Variable resistance case

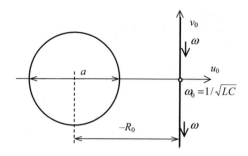

$$W^{-1}(j\omega) = \frac{LC(j\omega)^2 + 1}{C(j\omega)}.$$

The frequency characteristic coincides with the imaginary axis (see Fig. 14.4), and therefore the balance equation has no solution in the case of a positive resistance.

14.2   Harmonic Oscillation Generator

There are many designs of harmonic oscillation generators in various engineering fields. One of the simplest designs is built on the basis of a linear dynamic part (or oscillator) which is looped by a positive nonlinear elastic (capacitive) feedback. Let the description of the linear part be

$$W(p) = \frac{ks}{T_1 T_2 s^2 + T_2 s + 1}.$$

Let an oscillation amplitude limiter be realized as a nonlinear real two-position relay feedback. The description of such kind of relay was given by expression (4.24) and presented in Fig. 4.3. The harmonic linearization factor derived for those conditions are

$$W(A) = \frac{4B}{\pi A}(\cos \psi_1 - j \sin \psi_1), \psi_1 = \arcsin \frac{b}{A}. \tag{14.3}$$

14.2.1   Self-oscillation Evaluation

The conditions of self-oscillation existence follow from (4.32) provided that the negative sign is reversed owing to a positive feedback

$$W^{-1}(j\omega) = W(A). \tag{14.4}$$

Fig. 14.5 Generator.
Inversed Nyquist hodograph
plane

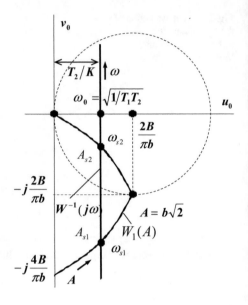

The inversed Nyquist hodograph

$$W^{-1}(j\omega) = \frac{T_2}{k} - j\frac{1 - T_1 T_2 \omega^2}{k\omega} \tag{14.5}$$

is the vertical line in the right half-plane at the distance of T_2/k from the imaginary axis (see Fig. 14.5). According to condition (14.4), the harmonic linearization factor hodograph $W_1(A)$ is contoured in the same half-plane. The hodograph has the maximum of its real part $2B/\pi b$ at point $A = \sqrt{2}b$. The hodograph cross points give the target amplitudes and frequencies of the self-oscillations.

As follows from Fig. 14.5, the self-oscillation existence condition for the frequencies $\omega^2 < 1/T_1 T_2$ is given by the inequality

$$\mathrm{Re}W^{-1}(j\omega) < \mathrm{Re}W(A = \sqrt{2}b)$$

or

$$k > k_{cr} = \pi\, T_2 b/2B. \tag{14.6}$$

Substituting characteristics (14.3) and (14.5) into condition (14.4), from the real part equality

$$1 - \frac{b^2}{A^2} = \frac{\pi^2 A^2 T_2^2}{16k^2 B^2}$$

the relative generator amplitude is found as

$$a^2 = \frac{A^2}{b^2} = \frac{1 \pm \sqrt{1 - \beta^2}}{0,5\beta^2}; \quad \beta = \frac{\pi b T_2}{2Bk} < 1. \tag{14.7}$$

From the imaginary part equality

$$\frac{4Bk}{\pi b} \frac{b^2}{A^2} = \frac{1 - \omega^2 T_1 T_2}{\omega}; \quad \omega < \omega_0 = \sqrt{\frac{1}{T_1 T_2}}$$

the generating frequency is found in the form

$$\omega_a = -\frac{1}{\beta a^2 T_1} + \sqrt{\frac{1}{\beta^2 a^4 T_1^2} + \frac{1}{T_1 T_2}}. \tag{14.8}$$

14.2.2 Generation Stability

As appeared in Fig. 14.5, these two solutions correspond to the two hodograph cross points and they are the self-oscillation amplitudes, A_{s1}, and, A_{s2}.

It follows directly from Fig. 14.5 that the derivative $d\,\mathrm{Re}\,W(A)/dA$ is positive as long as $b < A < \sqrt{2}b$ and it is negative if $\sqrt{2}b < A < \infty$. This means the center of the first parametric resonance circle is to the right from the vertical line $u_0 = T_2/k$ at low amplitudes A, and to the left from that if the amplitudes $A > \sqrt{2}b$. Since the unstable region is located to the right from the inversed Nyquist hodograph and the stable one is to the left from it, small self-oscillations (A_{s1}, ω_{s1}) are unstable, whereas the large oscillations (A_{s2}, ω_{s2}) are stable. This is also readily confirmable from the consideration of the small increments for points $A_s \pm \Delta A$, as stated in Part 3. Thus, the necessary stability condition of self-excited oscillations in the generator has been found "on average".

Let us understand if the generator oscillation stability loss is possible because of the excitation of high-harmonic parametric resonances. Since the circle radiuses of the high-harmonic parametric resonances are equal to those for the first parametric resonance, according to Table 4.1 for $K_p = B$

$$r_k = |\rho_1| = \frac{2B}{\pi A \sqrt{1 - \frac{b^2}{A^2}}} = \frac{2B}{\pi \sqrt{A^2 - b^2}}.$$

As followed from Fig. 14.5, the lowest generated frequency is feasible at $A = \sqrt{2}b$ and the corresponding circle radius is $r_k = \frac{2B}{\pi b}$. All the circles coincide and cross the origin and the point $A = \sqrt{2}b$ of the hodograph $W(A)$, as also shown in Fig. 14.5 by the dotted line. The circle-enclosed limit frequencies are derived from the equation

$$\mathrm{Im}\, W^{-1}(j\omega) = \pm \frac{2B}{\pi b}$$

in the form

$$\frac{1 - \omega^2 T_1 T_2}{\omega k} = \pm \frac{2B}{\pi b}.$$

This implies

$$\omega_{1,2} = \frac{\sqrt{\lambda^2 k^2 + 4T_1 T_2} \pm \lambda k}{2T_1 T_2}, \; \lambda = \frac{2B}{\pi b}.$$

The high-frequency parametric oscillation excitation can occur, where just one of the upper odd harmonic frequencies get to the circle. Since the third harmonic is the nearest one, at least the inequality

$$\omega_1 > 3\omega_2,$$

has to be met, that is to say,

$$k > \frac{\pi b}{\sqrt{3}} \frac{\sqrt{T_1 T_2}}{B}.$$

or in critical gain coefficient denotation (14.6)

$$k > \frac{2k_{cr}}{\sqrt{3}} \sqrt{\frac{T_1}{T_2}}$$

Hence, the stability loss condition for the generator oscillations in its critical case $k = k_{cr}$ are

$$\frac{2}{\sqrt{3}} \sqrt{\frac{T_1}{T_2}} < 1$$

or

$$\xi = \sqrt{\frac{T_1}{T_2}} < \frac{\sqrt{3}}{2} \quad or \quad T_1 < 0.75 T_2.$$

The parametric resonance is excited at higher harmonics.

Following the similar reasoning but for arbitrary oscillation amplitude (12.4), the general oscillation stability loss condition can be derived as

$$\frac{T_1}{T_2} = \xi^2 < \frac{3}{4} \frac{\beta^2}{(1 + \sqrt{1 - \beta^2})^2}, \; \beta = \frac{k_{cr}}{k} < 1.$$

Fig. 14.6 Generator.
Oscillatory processes

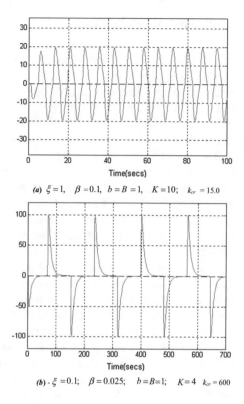

(a) $\xi = 1$, $\beta = 0.1$, $b = B = 1$, $K = 10$; $k_{cr} = 15.0$

(b) - $\xi = 0.1$; $\beta = 0.025$; $b = B = 1$; $K = 4$ $k_{cr} = 600$

It is interesting to note that the last condition is close to the boundary conditions for the describing function method applicability. Indeed, the linear part or oscillator does not satisfy the resonance condition as long as $\xi^2 = T_1/T_2 < 0.25$, and then the stability loss condition gives

$$0.87 < \beta < 1.$$

The excitation range seems to be too narrow and it questions the reliability of the method, especially knowing that it is approximate. Thus, the generator oscillation stability loss in the form of parametric oscillations excitation, specifically at the third harmonic frequency, happens when the describing function applicability conditions fail. It is significant that the stability loss takes place as a loss of a single-frequency oscillation profile.

A single-frequency oscillation takes place in the system in Fig. 14.6a, for $\xi = T_1 = b = B = 1$, $k_{cr} = 1.57$, $\beta = 0.1$ and Fig. 14.6b displays multifrequency motion with $T_1 = b = B = 1$, $\xi = 0.1$, $k_{cr} = 15$, $\beta = 0.25$; stability loss condition is met.

14.3 Rotating Converter[1]

Rotating electric energy converters (RCs) serve as classic examples of periodically variable parameter systems in electromechanics. These are motors and generators whose winding inductance depends on a rotary angle. In most cases (implicit-pole machines, high rotary speeds, slight tooth mesh effects, etc.), the periodic variation of winding inductance is managed to neglect but the operability of converters is frequently threatened with rotational speed-timed inductance variations.

The complex processes in rotary converters have attracted the attention of researchers for ages. The investigations were initiated in outstanding works performed by scientists of the Leningrad polytechnic school such as A. A. Gorev, M. P. Kostenko, A. I. Vazhnov, M. L. Levinshtein, etc. The analysis of the solutions of periodically unsteady equations concerning electric machines by converting those to constant coefficient equations is represented by well-known Gorev-Park equations.

In general way, the complex processes in electromechanical systems are described by nonlinear differential equations and have no exact closed-form solutions. The determination and analysis of the periodic motion stability in those systems by using asymptotic techniques and also qualitative methods relate to the class of fundamental problems and require outstanding mathematical qualification.

An autonomous power plant is considered to illustrate the first harmonic method efficiency therein. The plant consists of a direct current motor and a single-phase synchronous generator. It works on various lumped and distributed loads. The ship-borne DC–AC converters for accumulator batteries are typical examples for such electromechanical systems.

14.3.1 Description of Electromechanical Converter (EMC)

Let us consider the motor–generator system shown in Fig. 14.7.

The combined equations coupling the specified variables with each other in a single-phase converter are as follows:

$$
\begin{aligned}
J\ddot{\alpha} &= C_M I - M_r \\
U &= RI + L\dot{I} + C_E \omega \\
M_r &= \frac{\partial M}{\partial \alpha} i I_e \\
\frac{\partial M}{\partial t} I_e &= Z_n i,
\end{aligned}
\tag{14.9}
$$

where J is the inertia of the shaft and the rotors in the motor–generator system, α is the shaft rotation angle, C_M is the electromechanical motor constant, C_E is the

[1]Y. K. Lee (Republic of Korea) took part in the work [14]

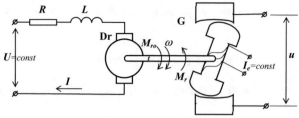

Designations:

U and I are the voltage and current in the direct current motor circuit, correspondingly; R and L are the active resistance and inductance of the direct current motor, correspondingly; M_{rot} and M_r are the motor torque and the generator antitorque moment, correspondingly;	ω is the angular velocity of the motor, the shaft and the generator; I_e is the excitation current of the generator armature; u is the load circuit voltage.

Fig. 14.7 Motor–generator system

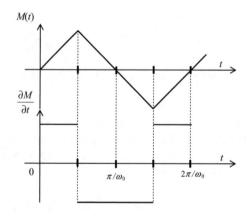

Fig. 14.8 Mutual inductance variation during anchor rotation

electric constant of the motor; M is the mutual inductance of the stator winding and the generator armature; Z_n is the complex impedance of a load; i is the stator current.

The relative relocation of the stator and armature windings which is stimulated by the armature rotation is reflected in the fact that M is needed to assume as a periodic time function. The idealized law of the time variation of M and $\partial M/\partial t$ is usually supposed under the constant shaft rotation speed, as shown in Fig. 14.8.

In real conditions, the harmonic approximation of the relation looks more reasonable in the form

$$M(t) = C_U \sin \alpha = C_U \sin \omega_0 t, \tag{14.10}$$

where C_U is a maximum value of the mutual inductance at the complete coincidence of winding exes.

Applying Laplace transformation to the system and passing from the combined equations to the single one in the shaft rotation angle, α, Eq. (14.10) takes the form

$$\begin{aligned}
T_a T_M s^3 \alpha + T_M s^2 \alpha + s\alpha + k_1 W_n(s) \left(\frac{\partial M}{\partial \alpha}\right)^2 s\alpha \\
+ T_a k_1 W_n(s) \left\{ 2\frac{\partial M}{\partial \alpha} \left(\frac{\partial^2 M}{\partial \alpha^2}\right) s^2 \alpha^2 + \left(\frac{\partial M}{\partial \alpha}\right)^2 s^2 \alpha \right\} = \frac{U}{C_E},
\end{aligned} \tag{14.11}$$

where

$$k_1 = \left(\frac{T_M I_e^2}{J}\right), \ T_a = \frac{L}{R}, \ T_M = \frac{RJ}{C_E C_M}$$

and $W_n(s) = \frac{1}{Z_n}$ is the voltage-to-current transfer function in the load circuit.

Since the mutual inductance M is the periodic function of a rotation angle, (14.11) is nonlinear. The steady motion of the rotating rotor with a constant speed, ω_0 is considered. To study motion stability, transferring to variation equations is performed by entering the small angular deviations $\Delta\alpha$ from the steady rotation $\alpha = \omega_0 t$. Then, (14.11) is written down in the form

$$\begin{aligned}
T_a T_M s^3 \Delta\alpha + T_M s^2 \Delta\alpha + s\Delta\alpha + \left\{ 2k_1 W_n(s) \left(\frac{\partial M}{\partial \alpha}\right) \left(\frac{\partial^2 M}{\partial \alpha^2}\right) \omega_0 \right\} \Delta\alpha \\
+ \left\{ k_1 W_n(s) \left(\frac{\partial M}{\partial \alpha}\right)^2 s \right\} \Delta\alpha + T_a k_1 W_n(s) \left[2 \left\{ \left(\frac{\partial^2 M}{\partial \alpha^2}\right)^2 \omega_0^2 + \left(\frac{\partial M}{\partial \alpha}\right) \left(\frac{\partial^3 M}{\partial \alpha^3}\right) \omega_0^2 \right. \right. \\
\left. \left. + \left(\frac{\partial M}{\partial \alpha}\right) \left(\frac{\partial^2 M}{\partial \alpha^2}\right) (2\omega_0) s \right\} + \left(\frac{\partial M}{\partial \alpha}\right)^2 s^{s2} \right] \Delta\alpha = 0
\end{aligned}$$

$$\tag{14.12}$$

14.3.2 Harmonic Stationarization and Parametric Oscillation Excitation Conditions

Let us group the terms of (14.12) with respect to the power of the operator s

$$
T_a T_M s^3 + \left\{ T_M + T_a k_1 W_n(s) \left(\frac{\partial M}{\partial \alpha} \right)^2 \right\} s^2
$$

$$
+ \left\{ 1 + k_1 W_n(s) \left(\frac{\partial M}{\partial \alpha} \right)^2 + 4 T_a \omega_0 k_1 W_n(s) \left(\frac{\partial M}{\partial \alpha} \right) \left(\frac{\partial^2 M}{\partial \alpha^2} \right) \right\} s
$$

$$
+ \left\{ 2 \omega_0 k_1 W_n(s) \left(\frac{\partial M}{\partial \alpha} \right) \left(\frac{\partial^2 M}{\partial \alpha^2} \right) + 2 T_a \omega_0^2 k_1 W_n(s) \left(\frac{\partial^2 M}{\partial \alpha^2} \right)^2 \right.
$$

$$
\left. + 2 T_a \omega_0^2 k_1 W_n(s) \left(\frac{\partial M}{\partial \alpha} \right) \left(\frac{\partial^3 M}{\partial \alpha^3} \right) \right\} = 0
$$

Since

$$
M = C_U \sin \alpha
$$

and correspondingly

$\frac{\partial M}{\partial \alpha} = C_U \cos \alpha, \ \frac{\partial^2 M}{\partial \alpha^2} = -C_U \sin \alpha$ и $\frac{\partial^3 M}{\partial \alpha^3} = -C_U \cos \alpha$,

the last equality is rewritten as follows:

$$
T_a T_M s^3 + \left(T_M + T_a k_1 W_n(s) C_U^2 \cos^2 \alpha \right) s^2
$$
$$
+ \left\{ 1 + k_1 W_n(s) C_U^2 \cos^2 \alpha - 4 \omega_0 T_a k_1 W_n(s) C_U^2 \cos \alpha \sin \alpha \right\} s
$$
$$
+ \left\{ -2 \omega_0 k_1 W_n(s) C_U^2 \cos \alpha \sin \alpha + 2 T_a \omega_0^2 k_1 W_n(s) C_U^2 \sin^2 \alpha \right.
$$
$$
\left. - 2 T_a \omega_0^2 k_1 W_n(s) C_U^2 \cos^2 \alpha \right\} = 0
$$

or

$$
T_a T_M s^3 + T_M s^2 + p + \left(T_a q s^2 + q s - 2 T_a \omega_0^2 q \right) \cos^2 \alpha
$$
$$
+ 2 T_a \omega_0^2 q \sin^2 \alpha - (4 \omega_0 T_a q s + 2 \omega_0 q) \cos \alpha \sin \alpha = 0,
$$

where $q = k_1 W_n(s) C_U^2$.
  Taking into account the trigonometric equalities

$$
\cos^2 \alpha = \frac{1 + \cos 2\alpha}{2}, \ \sin^2 \alpha = \frac{1 - \cos 2\alpha}{2}, \text{ and } \cos \alpha \sin \alpha = \frac{\sin 2\alpha}{2},
$$

the expression takes the form

$$T_a T_M s^3 + T_M s^2 + s + \left(T_a q s^2 + q s - 2 T_a \omega_0^2 q\right) \left(\frac{1 + \cos 2\alpha}{2}\right)$$
$$+ 2 T_a \omega_0^2 q \left(\frac{1 - \cos 2\alpha}{2}\right) - (4\omega_0 T_a q s + 2\omega_0 q) \left(\frac{\sin 2\alpha}{2}\right) = 0.$$

By virtue of $\alpha = \omega t$ is a time function, the stationarization conformably to the first parametric resonance results in the equation

$$T_a T_M s^3 + T_M s^2 + s + \frac{1}{2} q \left(T_a s^2 + s - 2 T_a \omega_0^2\right) \left(1 - \frac{1}{2} e^{-j\varphi}\right)$$
$$+ T_a \omega_0^2 q \left(1 + \frac{1}{2} e^{-j\varphi}\right) - \omega_0 q \left(2 T_a s + 1\right) \left(\frac{j}{2} e^{-j\varphi}\right) = 0,$$

where φ is the phase difference between the system oscillations and those of the parameter.

The equation takes the following form after pooling the items comprising exponential factors

$$T_a T_M s^3 + \left(T_M + \frac{1}{2} q T_a\right) s^2 + \left(1 + \frac{1}{2} q\right) s$$
$$+ \left\{\left(\frac{1}{2} q T_a\right) s^2 + \left(\frac{1}{2} q + j\omega_0 (2 T_a) q\right) s + j\omega_0 q\right\} \left(-\frac{1}{2} e^{-j\varphi}\right) = 0.$$

Whence the first parametric resonance excitation condition arises as

$$1 + W(s) W(j\varphi) = 0,$$

where

$$W(s)|_{s=j\frac{\Omega}{2}} = \frac{\left(\frac{1}{2} q T_a\right) s^2 + \left(\frac{1}{2} q + j\omega_0 (2 T_a) q\right) s + j\omega_0 q}{T_a T_M s^3 + \left(T_M + \frac{1}{2} q T_a\right) s^2 + \left(1 + \frac{1}{2} q\right) s}\bigg|_{s=j\frac{\Omega}{2}}$$
$$= \frac{\frac{5}{2} q T_a \left(j\frac{\Omega}{2}\right) + \frac{3}{2} q}{T_a T_M \left(j\frac{\Omega}{2}\right)^2 + \left(T_M + \frac{1}{2} q T_a\right) \left(j\frac{\Omega}{2}\right) + \left(1 + \frac{1}{2} q\right)},$$

$$W(j\varphi) = \left(-\frac{1}{2} e^{-j\varphi}\right),$$

and $\Omega = 2\omega_0$ is the parameter variation frequency. Therefore, the approximate first parametric resonance excitation condition is ensued as follows:

Fig. 14.9 Equivalent circuit
for derivation of first
parametric resonance
excitation boundaries

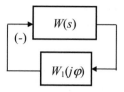

$$T_a T_M s^3 + \left(T_M + \frac{1}{2} T_a q \right) s^2 + \left(1 + \frac{1}{2} q \right) s$$
$$+ \left\{ \frac{1}{2} q T_a s^2 + \left(\frac{1}{2} q + s(2 T_a) q \right) s + s q \right\} \left(-\frac{1}{2} e^{-j\varphi} \right) = 0, \tag{14.13}$$

where $q = k_1 W_n(s) C_u^2$, here $s = j\omega_0 = j\Omega/2$, Ω is the parameter variation frequency and in the case of a single pole machine which is under consideration $\Omega = 2\omega_0$.

Equation (14.13) can be regarded as a characteristic polynomial with an exponential factor of some linear steady system. The condition when the left part of (14.13) is zero is equivalent to that, where the feedback system as shown in Fig. 14.9 is at its stability boundary in the agreed notations

$$W(j\Omega) = \frac{\frac{5}{2} T_a q \left(j\frac{\Omega}{2} \right) + \frac{3}{2} q}{T_a T_M \left(j\frac{\Omega}{2} \right)^2 + \left(T_M + \frac{1}{2} T_a q \right) \left(j\frac{\Omega}{2} \right) + \left(1 + \frac{1}{2} q \right)} \tag{14.14}$$

$$W_1 (j\varphi) = \left(-\frac{1}{2} e^{-j\varphi} \right). \tag{14.15}$$

Thus, while rotor–generator system (14.9) operates into the load circuit with transfer function $W_n(s)$ under mutual inductance variant law (14.10), according to Nyquist criterion the steady rotation of the shaft at the speed of ω_0 in that system **is unstable**, if the hodograph of the frequency response function $W(j\Omega/2)W_1(j\varphi)$ encircles the point $(-1; j0)$. The last statement also means that the critical parameter variation frequencies and amplitudes to observe the first parametric resonance excitation can be obtained from the condition

$$W(j\Omega/2)W_1(j\varphi) = -1 \quad \text{or} \quad W_1(j\varphi) = -W^{-1}(j\Omega/2). \tag{14.16}$$

Let us consider the parametric resonance excitation conditions with various types of the load, Z_n.

Active load. It should be assumed $W_n(s) = 1/r_n$ in the relations for q in conditions (14.13) and (14.14) as long as EMC is loaded by the active load, as shown in Fig. 14.10. In this particular case, the analytic expression can be derived to calculate the excitation range boundary of the first parametric resonance in the form

$$\Omega = \sqrt{\frac{2}{\mu_1} \left(-\mu_2 \pm \sqrt{\mu_2^2 - 4\mu_1\mu_3} \right)}, \tag{14.17}$$

Fig. 14.10 EMC with
lumped active load

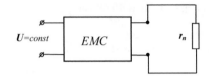

Fig. 14.11 EMC with
lumped reactive load

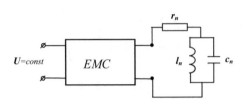

where

$$\mu_1 = (T_a T_M)^2,$$

$$\mu_2 = \left(T_M^2 - \frac{21}{16} \left(T_a k C_U^2 \right)^2 - 2 T_a T_M \right),$$

$$\mu_3 = \left(1 + k C_U^2 - \frac{5}{16} \left(k C_U^2 \right)^2 \right), k = \frac{k_1}{r_n}.$$

The parameter variation frequency Ω is a real value if Nyquist hodograph of the stationary part crosses the circle of stationarized component frequency response.

Reactive load. Let EMS have an electric oscillatory circuit as a load. The circuit includes the inductance l_n, the resistance r_n, and the capacitance c_n, as presented in Fig. 14.11.

In that case, it should be assigned

$$W_n(s) = \frac{1}{Z_n} = \frac{l_n c_n s^2 + 1}{l_n c_n r_n s^2 + l_n s + r_n} \tag{14.18}$$

in the relations for q in conditions (14.13) and (14.14). The first parametric resonance excitation condition follows from expressions (14.14) to (14.16) and (14.18) with $s = j\Omega/2$.

Distributed load. Modern huge ship or submarine power-supply systems a cable path length can reach kilometers. The cables compactly locate near each other in the insulation channels or twisted and therefore inevitably cross-talk. The phenomena essential for distributed parameter systems emerge. The distributed parameter systems are described by partial differential equations that are complex and rarely release the analytical solution in general form. But to engage the stationarization method, we are rather interested in a specific transfer function (or frequency response) from a certain input to certain output, rather than the general solution in time-space domain.

At present, the finite element method (FEM) is a basic technique used by mechanical engineers. Nevertheless, the method requires great computational capabilities in the high-precision calculations of dynamic problems, since it implies solving the problem for each node of the mesh. We are going to show, however, how to avoid this complexity for certain uniform objects at the expense of analytical efforts.

Thus, a long electric line extends along the x-axis and has the following model

$$\frac{\partial u}{\partial x} = -r_L i - l_L \frac{\partial i}{\partial t}$$
$$\frac{\partial i}{\partial x} = -c_L \frac{\partial u}{\partial t},$$
(14.19)

where $u(x,t)$ and $i(x,t)$ are the voltage and current in the circuit, correspondingly; r_L, l_L, and c_L are the resistance, inductance, and capacitance per length unit of the long line, correspondingly. Boundary conditions are to complete Eq. (14.19).

To get the long line frequency characteristic, we plan to find the relation between the current and voltage at the beginning and end points of the line under zero initial conditions.

Let us focus on the values $u(x)$ at the equal-spaced points $u(n)$, $n = 0, \dots N$ with the interval $h = l/N$ only. An operator of coordinate differentiation is approximated by a bilinear finite-difference scheme and Laplace transform is applied to time differentiation operator. Thereafter system (14.19) appears in the form

$$\frac{2}{h}\{u(n+1) - u(n)\} = -(r_L + l_L s)\{i(n+1) + i(n)\}$$
$$\frac{2}{h}\{i(n+1) - i(n)\} = -c_L s\{u(n+1) + u(n)\} \qquad n = 0, \dots, N-1, \quad (14.20)$$

where s is Laplace operator and h is the finite element length of the line.

The system of algebraic equations which can be solved with respect to the voltage and current is derived by applying z-transformation to (14.20)

$$U(z) = \frac{(z^2 - z \cosh \xi)u(0) - \beta \cdot zi(0)}{z^2 - 2z \cosh \xi + 1}$$
$$I(z) = \frac{(z^2 - z \cosh \xi)i(0) - \gamma \cdot zu(0)}{z^2 - 2z \cosh \xi + 1},$$
(14.21)

where

$$\cosh \xi = \frac{\left(\frac{2}{h}\right)^2 + c_L s(r_L + l_L s)}{\left(\frac{2}{h}\right)^2 - c_L s(r_L + l_L s)},$$

$$\beta = \frac{\left(\frac{4}{h}\right)(r_L + l_L s)}{\left(\frac{2}{h}\right)^2 - c_L s(r_L + l_L s)},$$

$$\gamma = \frac{\left(\frac{4}{h}\right) c_L s}{\left(\frac{2}{h}\right)^2 - c_L s(r_L + l_L s)}.$$

Using z-transformation table for elementary functions, the relationship can be obtained with respect to origins as the function of the point number

$$
\begin{bmatrix} u(n) \\ i(n) \end{bmatrix} = \begin{bmatrix} \cosh \xi n & -\beta \cdot \frac{\sinh \xi n}{\sinh \xi} \\ -\gamma \cdot \frac{\sinh \xi n}{\sinh \xi} & \cosh \xi n \end{bmatrix} \begin{bmatrix} u(0) \\ i(0) \end{bmatrix}.
$$

The voltage signs corresponding to the edge conditions at the left and right ends have to be different since the line current in the chosen direction charges, for example, a capacitance at the left end and discharges the capacitance when it is located at the left one. With regard to the above mentioned, the relations between the current and voltage values at either of the line ends are rewritten in the form

$$
\begin{bmatrix} u_0 \\ u_N \end{bmatrix} = \frac{\sinh \xi}{\gamma} \begin{bmatrix} \frac{1}{\tanh \xi N} & -\frac{1}{\sinh \xi N} \\ -\frac{1}{\sinh \xi N} & \frac{1}{\tanh \xi N} \end{bmatrix} \begin{bmatrix} i_0 \\ i_N \end{bmatrix}. \tag{14.22}
$$

with agreed notations $i(0) = i_0$, $u(0) = u_0$, $i(N) = i_N$ and $u(N) = -u_N$.

If there is a reactive load $u_N = -R(s) \, i_N$ at the end of the line, the communication between the ends follows from (14.22) as

$$
u_0 = \left[\frac{\sinh \xi}{\gamma} \sinh \xi N + \cosh \xi N \cdot R(s) \right] i_N = \frac{1}{W_n(s)} i_N.
$$

Hence, when there is a reactive load $R(s)$ at the end of the long line, the transfer function of the load $W_n(s)$ has the finite-dimensional approximation

$$
W_n(s) = \frac{1}{\frac{\sinh \xi}{\gamma} \sinh \xi N + \cosh \xi N \cdot R(s)}, \tag{14.23}
$$

where N is the amount of cells in the finite-dimensional model, ξ and γ are some parameters dependent on the operator s, the physical parameters of a unit length, and an approximation order chosen. With a purely active load r_n at the end of the long line $R(s) = r_n$; $R(s) = l_n s$ in the case of the pure inductance l_n. The reactive load in the form of the oscillatory circuit as illustrated in Fig. 14.10 gives

$$
R(s) = \frac{l_n c_n r_n s^2 + l_n s + r_n}{l_n c_n s^2 + 1}.
$$

At $N = 0$, in the last case $W_n(s)$ is exactly the same as the lumped load transfer function (14.18).

14.3.3 Excitation Condition Calculation and Numerical Simulation of Parametric Resonance

The physical parameters for numerical examples were selected to be realistic. Thus, a direct current motor was taken as a major device with performance: the nominal rotation frequency $\omega_0 = 3000$ rpm, the power of 5.5 kW, the supply voltage $U = 440$ V. The technical characteristics of the generator were: the voltage $u = 230$ V, the power of 4 kW, the exciting armature current $I_{exc} = 3$A. The total shaft inertia J is assumed to be 0.029 kGm^2.

Active load. Let us consider the active load with resistance $r_n = 10\ \Omega$. Using $W_n(s) = 1/r_n$, in the relation for q in (14.13), the final form of the stationary part hodograph $W(s)$ is found from (14.13) and (14.14). The hodographs $-W_1^{-1}(j\varphi)$ and $W(j\Omega/2)$ are shown together in Fig. 14.12a.

Since there is no crossing of the hodographs and the whole hodograph $W(j\Omega/2)$ sits inside the circle $-W_{1r}^{-1}(j\varphi)$, the first parametric resonance is unfeasible. Figure 14.12b shows the calculation result to the same hodographs under the strongly lowered load $r_n = 0.1\ \Omega$. In this case, instability is possible at the rotating velocities from zero to 202.98 rad/s, whereas the parameter variation frequency at a regular mode is coincident with the rated rotating velocity of the shaft and equal to 300 rad/s. Hence, the parametric resonance is theoretically feasible with a small active resistance but in practice, such kinds of electric systems do not exist.

Reactive load. Just now, let us consider the lumped purely inductive load $l_n = 0.032\ H$, while $W_n(s) = l_n s$. Figure 14.13 presents the hodographs $-W_1^{-1}(j\varphi)$ and $W(j\Omega/2)$.

As followed in Fig. 14.13, the first parametric resonance is accessible when the rotary speeds are lower than 29.908 rad/s, which does not correspond to the rated duty of *EMC*.

Distributed load. Let us consider the operation of *EMC* into the 250 m line with the following parameters: $r_L = 1e{-}6\ \Omega/m$, $l_L = 1e{-}5\ H/m$, $c_L = 1e{-}8\ F/m$.

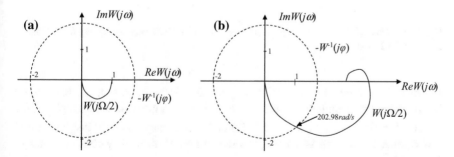

Fig. 14.12 On evaluation of the first parametric resonance excitation conditions. Lumped active load case. $-W_1^{-1}(j\varphi)$ is the stationarized element characteristic, $W(j\Omega/2)$ is frequency response of stationary part

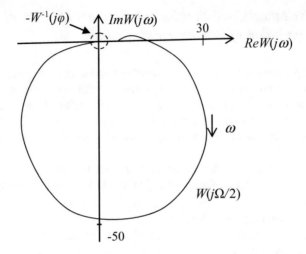

Fig. 14.13 On evaluation of the first parametric resonance excitation conditions. Lumped reactive load case

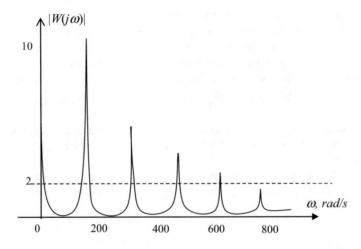

Fig. 14.14 On evaluation of the first parametric resonance, EMC operates on 250 m long line. **a** Hodographs, **b** frequency response

Let there be a lumped inductive load $l_n = 0.032$ H at the end of the line. $W_n(s)$ is calculated by formulation (14.23) then, Fig. 14.14a gives the hodographs of the transfer functions $-W_1^{-1}(j\varphi)$ и $W(j\Omega/2)$ in the frequency range limited by the first few resonances. Figure 14.14b depicts the amplitude–frequency characteristics of those transfer functions.

The calculation reveals the first parametric resonance is feasible in the vicinity of the first resonance frequencies: to 5.91, 151.06–151.49, 302.99–303.19,

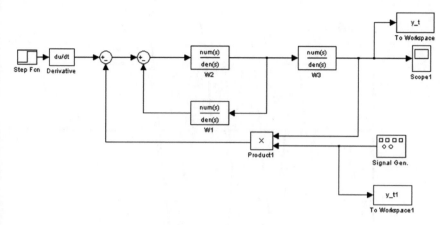

Fig. 14.15 Simulink scheme for simulation experiment

455.20–455.30 rad/s. The third resonance is the most interesting because it is in immediate proximity to the nominal rotation rate 300 rad/s.

Standard solvers are used for simulations. The setting is given in Simulink scheme (see Fig. 14.15), where the transfer functions $W_1(s)$, $W_2(s)$ are

$$W_1(s) = \frac{\frac{1}{2}k_1 C_U^2 (T_a s + 1)}{T_a T_M s^2 + T_M s + 1},$$

$$W_3(s) = \frac{\frac{1}{2}k_1 C_U^2 (5T_a s + 3)}{T_a T_M s^2 + T_M s + 1},$$

and the transfer function $W_2(s)$ is the load transfer function. In this case,

$$W_2(s) = W_n(s) = \frac{1}{\left[\frac{\sinh \xi}{\gamma} \sinh \xi N + \cosh \xi N \cdot R(s)\right]}$$

It is easy to agree that according to Fig. 14.7

$$W(s) = \frac{W_2(s)}{1 + W_1(s)W_2(s)} W_3(s)$$

and coincides with above used form (14.14).

Standard Simulink interface does not help when we the load transfer functions are not fraction or polynomials. However, there is no need to mimic the transfer function (14.23) exactly to witness the single-frequency parametric resonance picture. It is sufficient to take the transfer function approximation within the frequency range which is of interest.

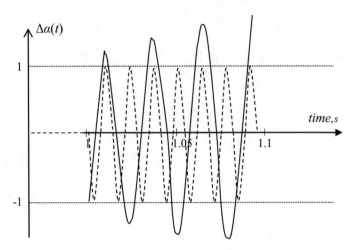

Fig. 14.16 Oscillogram of *EMC* rotor swing rise (fragment). Solid line represents coordinate oscillations, dashed line is parameter variation

The experiment revealed the presence of the first parametric resonance as shown in Fig. 14.16.

Chapter 15
Economics

15.1 Simulation Features in Macroeconomic Dynamics

The progress logic of mathematical modeling in economics was similar to any other field of engineering, e.g., mechanics. There was the evolution from simplest models to complex ones, from static relations to dynamic, from scalar problems to multidimensional, from linear problems to nonlinear, etc. However, the progress in classical mechanics was inseparably linked with the progress in mathematics and vice versa, they started applying well-developed by that time mathematical formalism to economy in the second half of the 20th century only. In many aspects, the difficulties of modeling of economics as a nature object are explained. Up to now, there is no such set of universally recognized laws which could be widely applicable, for example, Hook's law or Coulomb's law.

We anxiously state that the first attempt to use mathematical methods for the modeling of economic entities seem to have taken place several centuries ago. A. Cournot was among the first who employed differential calculus in 1838. Owing to that tool Valréas (1874) and Pareto (1908) enunciated the philosophy of overall economic balance which was further developed by Hicks and Samuelson later on. In the same period, the method and models of output material balance (introduced in 1930s by State Planning Committee (Gosplan) of the USSR) and the apparatus of production functions (Cobb and Douglas in 1920s) were developed and successfully applied in economic analysis. As the automatic control theory apparatus progressed, some ideas were applied for economic studies, for example, by J. Forrester. In the second half of the 20th century, the latest divisions of mathematics such as convex analysis, topology, theory of catastrophes and chaos, etc., found their applications in economics, see for example Zang's book appeared in 1999.

As in technical fields, economical dynamics describes the interconnection of basic parameters and variables by means of differential or finite-difference equations. These are usually linear time invariant equations or sometimes nonlinear differential ones. If the problem in focus is not only the analysis but also the management of

© Springer International Publishing AG, part of Springer Nature 2017
L. Chechurin and S. Chechurin, *Physical Fundamentals of Oscillations*,
https://doi.org/10.1007/978-3-319-75154-2_15

the economy, the theory of nonautonomous, that is, nonhomogeneous differential equations are used, including the apparatus of the theory of optimal control.

It is appropriate to rephrase the question of where dynamic models can be used in economics as follows: are there any economic processes in which the impact of the previous history on the behavior of a variable under consideration might be expected? In other words, what kinds of economic processes have the memory? It seems obvious that, for example, the prices and shopping behaviors in the market (in its highly concentrated form like stock or currency market) are not only influenced by their momentary values, but also the trends and history of selling. It also seems to be obvious that the economy itself could demand dynamic models for its description: the relation of the current economy state with the history seems yet clearer. The chapter focuses on dynamic models and attempts to design and analyze them.

We consider several models of oscillatory behavior in economic systems. Generally speaking, recognition of various cycles is a popular research topic in the economic science. Such are the qualitative theory of cyclic capitalism crisis by K. Marks, the theories of short "innovative waves" by J. Schumpeter or long cycles of current economic by N. Kondratiev. The theory of latter "waves" had some statistical background of long-term macroeconomic records of the developed countries.

The chapter uses some "physical" models of the economy. These are the models that take the form of differential equations. Certainly, basic linear differential equations are likely less equivalent instrument to describe actual processes in the economy in contrast to, say, how the second-order liner differential equation can describe small spring pendulum oscillations. First, since the economical system parameters (e.g., depreciation) are much harder to assess in contrast to those in physical systems. Second, both the parameters and all the structure of equations are known to be time-dependent to a certain extent. Third, even the laws themselves underlying the equations not always reflect actual variable interactions at the basic level. However, the models enable qualitative analysis of phase trajectories, the conditions of stability and instability in the economy, and therefore add to our understanding the economy dynamics.

15.2 Parametric Resonance in Dynamic Multisector Economy Balance Model

Let us consider the economic system of two business entities bearing some technological and financial ties, as shown in Fig. 15.1.

Each sector is described by its capital or the value of the fixed assets $K_i(t)$. It is assumed the entities are not able to produce anything alone. We also introduce the capital depreciation at the rate μ. Thus, the entity's capitals vanish in time as $\exp(-\mu t)$ if there is no inter sector cross–talk. This model is widely used in economics to describe elementary asset dynamics or in the Leontjev input–output model accurate with variables.

Fig. 15.1 Two-sector
economic system model

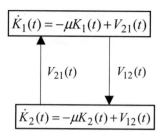

If the entities have common management, their assets can be moved from one entity to another according to the strategy chosen by the governing body. The question arises if there are any special asset passage modes to lead to instability in the system.

Let us model the cross-sector transactions as "taxes" from Sector 1 to Sector 2. Let us also specify that two types of "taxes" exist: stationary and time-varying. The stationary transactions reflect, for example, technological links between the two industries when the product of the first industry is used in the production process of the second one. The time-variant ties could mean a tax of variable rate. This controllable variation of the tax would be the control input to the system. So, the model of two equations appears as

$$\dot{K}_1 = -(\mu + \beta)K_1 + \alpha(t)K_2$$
$$\dot{K}_2 = -(\mu + \alpha(t))K_2(t) + \beta K_1, \tag{15.1}$$

where α is the variable asset/product share K_1, which is injected into the assets of industry two K_2, and $\beta(t)$ is the variable asset/product share of industry two K_2, which is injected back to the assets of industry one K_1.

It is not difficult to observe that the balance in the model is always kept, namely, the total assets decrease proportionally $\exp(-\mu t)$ as a result of summing up two equations in (15.1).

If the control action is periodical (for example, in case of rushing of investors from one industry to another), one will deal with the system of differential equations with a periodically variable coefficient. The parametric resonance can be expected to arise in the system at the specific frequency of a parameter (coefficient) variation.

Let us consider, the case when β is a constant coefficient and α, is a periodically variable function. For example,

$$a(t) = A \sin(\Omega t), \tag{15.2}$$

or

$$\alpha(t) = A \, \text{sgn}(\sin(\Omega t)), A < 1. \tag{15.3}$$

Fig. 15.2 Block diagram
with singled out time-variant
element in the feedback

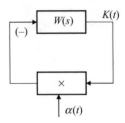

We use the stationarization method described in Part 2 to derive the solution. Let
us remind briefly the details of the method. First, system (15.1) has to be reduced to
one second-order equation, then Laplace transform is to be used. Stability analysis
of system (15.1) is equivalent to that for the feedback system given in Fig. 15.2. The
characteristic equation of that system is

$$K(t) + W(s)[\alpha(s)K(t)] = 0, \qquad (15.4)$$

where $K(t)$ is assumed to be the variable of the second-order equation derived by the
unification of two first-order from (15.1).

We replace the nonstationary part in (15.4) by its stationary harmonic approximation using Fourier-series expansion of the function $\alpha(t)$,

$$\alpha(t) = \sum_{r=-\infty}^{r=+\infty} \alpha_r e^{+j\Omega rt}, \Omega = \frac{2\pi}{T}, \alpha_r = \frac{1}{T} \int_0^T \alpha(t)e^{-j\Omega rt}dt.$$

The coefficients α_r are the gains of each specific harmonic

$$W^r(j\varphi) = \alpha_0 - \alpha_r e^{+jr\varphi}, \varphi = \Omega\tau, \quad r = 1, 2 \ldots,$$

here, φ is the phase difference between $\alpha(t)$ и $K(t)$; $r = 1$ as applied to the first
parametric resonance, i.e.,

$$W^1(j\varphi) = \alpha_0 - \alpha_1 e^{+j\varphi}. \qquad (15.5)$$

The coefficients α_r, for example, for the sinusoidal parameter variation can be
obtained as

$$\alpha_0 = 0, \alpha_1 = -\frac{jA}{2}, W_1(j\varphi) = \frac{jA}{2}e^{+j\varphi}. \qquad (15.6)$$

The hodograph of such a transfer function in the complex plane is the circle with
a center at the origin and a radius of $A/2$.

We can engage Nyquist stability criterion now. The boundary condition of the
system stability is

Fig. 15.3 Stationarized
system

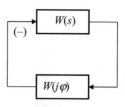

$$1 + W\left(j\frac{\Omega}{2}\right) W_{1\Gamma}(j\varphi) = 0,$$

which is equivalent to

$$-W_1(j\varphi) = W^{-1}\left(j\frac{\Omega}{2}\right). \tag{15.7}$$

This graphically means that if the inverse transfer response of the stationary part crosses the circle of the first stationary approximation (15.7), the first parametric resonance can take place. Moreover, the parameter variation frequencies are found based on the stationary hodograph part lying inside circle (15.7).

Example 1[1]. Let us consider two-agent economic system (15.1) in which α is the harmonic function $\alpha_t = A\sin(\Omega t)$. Let other parameters be

$$A = 0.7, \mu_1 = 0.2, \mu_2 = 0.1, \beta = 0.3, \tag{15.8}$$

i.e., the depreciation coefficients in Industry 1 and Industry 2 are 20 and 10% per year, correspondingly. Industry 1 passes permanently 30% of its assets to Branch 2. Branch 2 sometimes delegates, sometimes receives the assets from Branch 1 with the frequency Ω rad/year. The asset transfer amplitude is 70% of the assets of Branch 2. Using these parameters, the stationarized element transfer function (see Fig. 15.3) is

$$W(j\omega) = \frac{(j\omega) + 0.2}{(j\omega)^2 + 0.6(j\omega) + 0.05},$$

where $\omega = \frac{\Omega}{2}$ is the system oscillation frequency, Ω is the parameter oscillation frequency.

According to (15.6) with parameters (15.8), the transfer function is

$$W_1(j\varphi) = 0.35 j e^{+j\varphi}.$$

The hodograph of the transfer function is the circle with a center at the point $(0; j0)$ and a radius of 0.35.

---

[1]The calculations were performed by I. M. Shakhov.

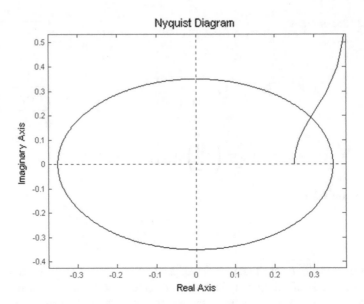

Fig. 15.4 On calculation of first parametric resonance excitation

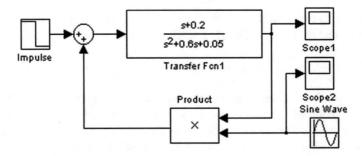

Fig. 15.5 Simulation setup for Simulink

The hodographs $-W_1(j\varphi)$ and $-W^{-1}(j\omega)$ cross (see Fig. 15.4). Thus, the first parametric resonance can be expected in the range of system frequencies 0–0.02 times/year, here and hereinafter, the frequencies are evaluated in the dimensions customized to economic processes.

The derived approximate parametric resonance excitation parameters can be verified in a numerical experiment. The Simulink software can be applied with a view to simulate the system oscillations under different parameter variation frequencies. Figure 15.5 illustrates the experiment scheme. Figure 15.6 presents the parametric resonance excitation patterns.

The numerical experiment gives the frequency range of the first parametric resonance excitation of [0…0.002] 1/year. The accuracy of the harmonic stationarization

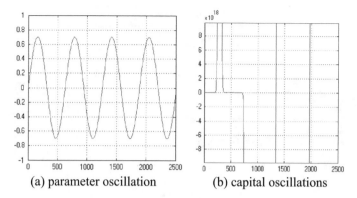

(a) parameter oscillation   (b) capital oscillations

Fig. 15.6 System oscillations

Fig. 15.7 Schematic of
three-sector economy

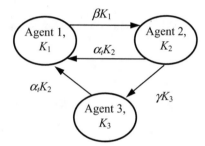

is of cause questionable in the case. However, the method provides qualitatively correct conclusion on system instability domain.

What seems to be hard to digest at the intuitive level is that a passive (i.e., non-productive) economic system demonstrate stability loss when we assume some elements of it to be time-variant. A situation like that can actually arise if the financial management has no long-term programs and rushes about their two or more businesses under the control. Financial manipulations could be even easier field to derive an example from. In all the cases it is expected although that if the enterprise is in debt (K is negative at certain time range), it borrows from outside.

As it is shown below the refined models based on Leontjev balance relations which include three or more associated institutions and take into account delays in asset development, the system oscillation frequencies can take more actual values.

15.2.1 Refined Models

Three-sector model. Another agent is introduced to extend the approach to a three-sector model (see Fig. 15.7). We consider new equation system (15.19) now.

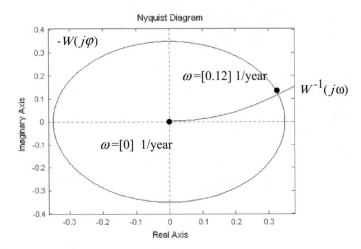

Fig. 15.8 On evaluation of excitation frequencies of first parametric resonance concerning three-sector model

$$\dot{K}_1 = -(\mu_1 + \beta)K_1 + 2\alpha_t K_2$$
$$\dot{K}_2 = \beta K_1 - (\mu_2 + \alpha_t)K_2 - \gamma K_3 \qquad (15.9)$$
$$\dot{K}_3 = -\alpha_t K_2 - \mu_3 K_3 + \gamma K_3.$$

The analytical approach coincides with the two-agent case, so let us pass on to an illustrative example at once. The following set of parameters are specified

$$\mu_1 = 0.1, \mu_2 = 0.05, \mu_3 = 0.2, \beta = 0.3, \gamma = 0.3.$$

As usual, the system is transformed to one equation of the third-order, Laplace transformation is applied, the nonstationary element is selected to the feedback, and the transfer function of a stationary unit is derived then as

$$W(j\omega) = \frac{(j\omega)^2 - 1.3j\omega - 0.38}{(j\omega)^3 + 0.35(j\omega)^2 - 0.025j\omega - 0.002}.$$

Using the harmonic linearization method, the approximate range of excitation frequencies to the first parametric resonance is obtained in the diagram form as shown in Fig. 15.8.

The parametric resonance can occur at the system oscillation frequencies $\omega = [0...0.12]$ 1/year, the parameter varies two times rarer, i.e., $[0...0.24]$ 1/year.

Distributed delay model. It was belied in our previous derivations that product and financial flows in the economic system products or finances are immediate. Obviously, it is far from to be correct in most cases. Even bank transfers require several days and it is needless to say how much longer the tangible transactions take. Another important voice of reality is that the transformation of financial assets into

Fig. 15.9 Two-commodity distributed delay model

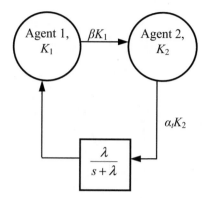

tangible capital or investment development takes some time. Classical models of economic dynamics can account all these transport, postal, or capitalization delays. We use the same approach and show how these delays influence the borders of the region of instability in the form of parametric resonance.

We first add one more equation to the system (15.9) (see system (15.10) and Fig. 15.9).

$$\dot{K}_1 = -(\mu_1 + \beta)K_1 + I$$
$$\dot{I} = -\lambda I + \lambda \alpha_t K_2 \qquad (15.10)$$
$$\dot{K}_2 = -(\alpha_t + \mu_2)K_2 + \beta K_1.$$

here, I is the amount of capital of industry 2 invested to the industry 1, λ is the delay parameter. What we see in (15.10) is well-known distributed lag model of the investment capitalization. The model is derived from the integral exponential kernel operator as

$$N(\theta) = \lambda e^{-\lambda \theta}$$

$$I(t) = \int\limits_0^\infty N(\theta)\alpha(t - \theta)X(t - \theta)d\theta.$$

Analogously to the previous stability evaluations, we transform the stationary part of the system to one higher-order differential equation, then apply Laplace transform and single out the nonstationary element as the feedback. Thus, the transfer functions of the stationary unit of $W_1(j\omega)$ and $W_2(j\omega)$ take the following form assuming certain lag parameters $\lambda = 0.1$:

$$W_1(j\omega) = \frac{(j\omega)^2 + 0.6j\omega + 0.02}{(j\omega)^3 + 0.7(j\omega)^2 + 0.11j\omega + 0.005},$$

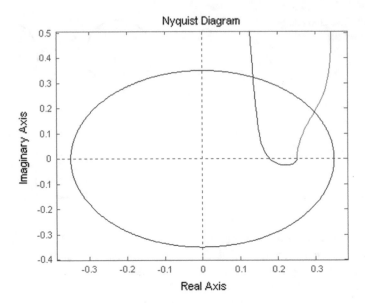

Fig. 15.10 On instability analysis for distributed lag model

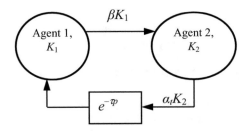

Fig. 15.11 Two-sector model with transport delay

$\lambda = 2$:

$$W_2(j\omega) = \frac{(j\omega)^2 + 2.5j\omega + 0.4}{(j\omega)^3 + 2.6(j\omega)^2 + 1.25j\omega + 0.1}.$$

As shown in Fig. 15.10, in the case of the system with the greater delay ($\lambda = 0.1$), the parametric resonance can arise at the oscillation frequencies $\omega = [0...0.06]$ 1/year and the system differs little from the two-agent one without a delay while $\omega = [0...0.02]$ 1/year.

Pure (transport) lag model. A transport delay is introduced to the economic two-sector system (see Fig. 15.11 and equation system (15.11)).

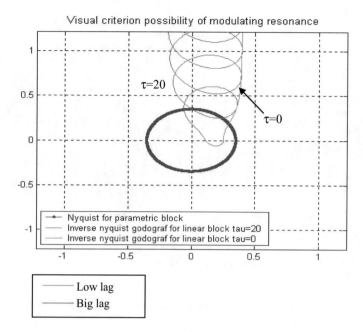

Fig. 15.12 On determination of instability parameters for pure delay model

$$\dot{K}_1 = -(\mu_1 + \beta)K_1 + I(t - \tau)$$
$$I = \alpha_t K_2 \qquad (15.11)$$
$$\dot{K}_2 = -(\alpha_t + \mu_2)K_2 + \beta K_1,$$

where τ is a pure (transport) time delay or time lag.

The stationarization method helps to define the instability range for this type of systems too. Let us assume the following set of parameters

$$\mu_1 = 0.2, \mu_2 = 0.1, \beta = 0.3.$$

In this case, the transfer function of the stationary part of the system is written in the form

$$W(s, \tau) = \frac{s + 0.5 - 0.3e^{-\tau s}}{s^2 + 0.6s + 0.05}.$$

Taking the delay into account changes the model parameters and affects the parametric resonance excitation regions (see Fig. 15.12). As long as the delay is big ($\tau = 20$ years), the range of "dangerous" frequencies becomes discontinuous.

We conclude, the paragraph by pointing out again that many assumptions we use seem unrealistic, even the idea of periodic investment redistribution. However, the aim is to demonstrate the possible influence of periodic parameter variation on the

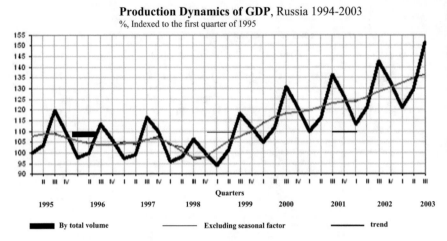

Fig. 15.13 Gross domestic product seasonal variations

stability of the economic models under consideration and a qualitatively new effect was found and demonstrated.

15.3 Harmonic Analysis of Goodwin Market Cycle Model

The hypothesizes of repeatability and cyclic nature of the events constituting the human history have been known, possibly, since Scriptural times. During the last century, they took the forms of scientific hypothesizes made on the basis of extensive statistical data. The works by A. L. Chizhevsky on the relation between periodic solar activity bursts and terrestrial diseases and epidemics (exogenous periodicity)[2] are striking examples. It is also worth to denote the theory of *ethnogenesis* by L. Gumilev in view of cyclical social structure changes (endogenous periodicity).

With the process of standardization of national economic accounts, many scientists turned to searching certain cycles in economic data series.[3] So J. Chumpeter offered the innovation cycling mechanism, N. Kondratiev introduced "long" macroeconomic indicator cycles, J. Kitchin investigated short periodic waves with a period of 2–4 years on the basis of the study of financial accounts and sale prices during a stock turnover, etc. Economic science knows hundreds of cycling theories.

Thus, if the theory of economic growth investigates the factors and conditions of the growth as a strictly directed long-term tendency, the economic cycle theory

---

[2]The title of the doctoral thesis of A. L. Chizhevsky "The investigation of world-historic process cycling" (Moscow 1920) speaks for itself.

[3]Figure 15.13 gives the elementary illustration of the above stated. The cycling easily explicable by a seasonal factor can be observed from the dynamics of Gross Domestic Product (GDP) for the past few years.

studies the causes of time-varying macroeconomic indicators, that is to say, their temporal progress rates.

This paragraph considers Goodwin market cycle model describing the comparable variation of the two economic indicators: the GDP consumption share and the employment share of the total population.

Goodwin model relies on the assumption that capital intensity, population growth rates, and labor production are constant. Another assumption is that the employment linearly depends on a wage growth rate.

There are two endogenous variables in the models: λ is the employment in the total population and δ is the consumption fund share of GDP. The variable dynamics is described in Goodwin model by the following system of equations:

$$\begin{cases} \dfrac{d\delta}{dt} = (a\lambda - a_0)\delta, \\ \dfrac{d\lambda}{dt} = (-b\delta + b_0)\lambda, \end{cases} \tag{15.12}$$

where a, a_0, b, b_0 are the model parameters dependent in a certain way on capital intensity, population increase rate, working efficiency rate, and coefficients describing the linear dependence of the employment on the wage rate.

Introduced model (15.12) is not linear and there are products of the unknown quantities on the rights.

Mathematics classifies the equations of (15.12) type as Lotka–Volterra's system. Similar equations and modifications thereof are used in various seemingly distant fields like multimode laser operations or chemical reactions and processes running in nuclear piles.

Let us consider, in more details, the biological area in which the model can also be applied to describe the joint changes of two populations. Let the populations be called "predators" and "preys". The predators can exist only in the presence of the preys. The preys reproduce themselves exponentially, that is to say, under the fixed growth rate in the condition when there are no predators (unlimited growth). If δ and λ are assumed to be the amounts of the predators and the preys, correspondingly, the derived Goodwin model reflects just the kind of "relationships". Indeed, the term $\delta\lambda$ reflects the coexistence and has a positive influence on the growth rate of the predators to be marked with a plus sign on the right part of the first equation. On the other hand, the term has a negative effect on a growth rate of the preys to reflect "the coexistence". The exponential degeneracy of the predators and the same reproduction the victims would be observed if there is no such a term. Such an analogy of (15.12) helps to understand the essence of variable relations.

The detailed analysis of similar models became the inevitable part of textbooks on biological population growth modeling or nonlinear oscillation theory. We briefly reproduce the analysis of the system here in the relevant to our objectives form.

The following equality comes from the dividing the first equation by the second

Fig. 15.14 Market cycle
phase profile

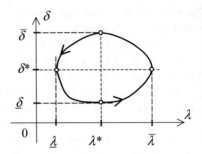

$$\left(-b + \frac{b_0}{\delta}\right) d\delta = \left(a - \frac{a_0}{\delta}\right) d\lambda. \tag{15.13}$$

Taking integrals of the left and right sides of (15.13) give

$$-b\delta + b_0 \ln \delta = a\lambda - a_0 \ln \lambda + \ln C$$

or

$$\delta^{b_0} e^{-b\delta} = C\lambda^{-a_0} e^{a\lambda}, \tag{15.14}$$

where C is a constant of integration.

The function $f(\delta) = \delta^{b_0} e^{-b\delta}$ has its maximum at the point $\delta^* = \frac{b_0}{b}$ and the function $g(\lambda) = C\lambda^{-a_0} e^{a\lambda}$ has its minimum at the point $\lambda^* = \frac{a_0}{a}$.

Moreover, as long as the argument is not negative, the left side of equality (15.14) increases from zero to maximum at $\delta = \delta^*$, and then decreases to zero. The right side diminishes from infinity to minimum at $\lambda = \lambda^*$, and then extends to infinity again. Thus, in the general case, the functions will have two cross points while there are sufficiently small values of C, that is to say, two values of λ can be found to any value of $\delta \neq \delta^*$ and vice versa. This means that the closed convex curve or cycle takes place in the plane (λ, δ) as shown in Fig. 15.14.

The same path patterns may be conceived by directly calculating the velocity flow in the phase plane based on the right sides of Eq. (15.12).

Since, the initial conditions can be basically arbitrary, an immense variety of the cycles can be obtained. This means that we deal with a conservative system. The system was used to describe an oscillation in a physical system, it would have meant that the system neither absorbs nor emits any energy over a cycle. It means, unfortunately, that the suggested model might correctly reflect the oscillatory nature of the introduced variables but not robust (according to the definition of differential equation system robustness by A. Andronov). Namely, an arbitrary small variation of system parameters as well as initial conditions result in a new cycle, that is, the oscillations with other amplitudes and frequencies. And even worse, small variation of the equation structure, for example, by introducing a quadratic term, can modify qualitatively the system phase flow.

The equilibrium point or the center of all the cycles (δ_E, λ_E) of such a system is easily found. If the system is at the equilibrium point, its derivatives are zero so the left sides of equation system (15.12) are also zero. Thus,

$$\delta_E = b_0/b = \delta^*,$$
$$\lambda_E = a_0/a = \lambda^*. \tag{15.15}$$

The latter implies that the semiaxis of all the cycles are parallel to the coordinate axis and cross the equilibrium point.

The cyclic processes have been detected in the system, but the process can not be given a closed mathematical description. Hence, we will try to derive the key parameters of the oscillations approximately. The above analysis revealed the absence of limit cycles in the system, so there is no sense in finding the oscillation amplitudes: the processes are completely defined by the initial conditions. It is interesting to evaluate the cycle frequency, at least approximately. Linearization is legitimate in the case of small oscillations near the equilibrium. For that, the solutions are searched in the form

$$\lambda = \lambda^* + \Delta_\lambda, \delta = \delta^* + \Delta_\delta,$$

where

$\Delta_\lambda, \Delta_\delta$ are small values. Substituting the last expressions into Eq. (15.12) and retaining the equal order terms, the following system of linear equations is derived as

$$\begin{cases} \dot{\Delta}_\delta = a\frac{b_0}{b}\Delta_\lambda \\ \dot{\Delta}_\lambda = -b\frac{a_0}{a}\Delta_\delta. \end{cases}$$

This system has the characteristic polynomial

$$s^2 + a_0b_0 = 0$$

and obviously pure imaginary roots as long as a_0b_0 are positive, which means the point (λ, δ) is the center. Hence, the market cycle frequency is

$$\omega = \sqrt{a_0b_0}$$

and the consumption oscillations in GDP has a $\pi/2$ phase-shift relative to the employment oscillations.

The origin is another equilibrium point in system (15.12). It is not difficult to observe that point is the saddle. The coordinate axes are separatrixs, the 0δ-axis is included in the saddle. The 0δ-axis originates from the latter.

Harmonic analysis as the first approximation. Let us return to the describing equations (15.12) and try to find a solution in the form of the harmonic functions

Fig. 15.15 Block diagram representation of Goodwin model

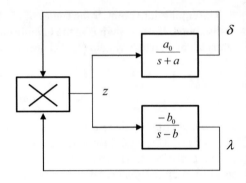

$$\delta\,(t) = \delta_0 + A\,\sin\omega t = \delta_0 + \delta_\sim,$$

$$\lambda\,(t) = \lambda_0 + B\,\sin(\omega t + \psi) = \lambda_0 + \lambda_\sim$$

where $\lambda_0\,\delta_0$ are the constant components of the oscillations at frequency ω; A and B are oscillation amplitudes, ψ is a phase shift between the oscillations of the coordinates: employment rate and the share of the consumption in GDP.

Let us introduce the additional variable z as

$$z = \lambda\delta.$$

Replacing the differentiation operator in system (15.12) by Laplace operator, we arrive at the following expression:

$$\begin{cases} z(s)\frac{a}{s+a_0} = \delta(s) \\ z(s)\frac{-b}{s-b_0} = \lambda(s). \end{cases} \tag{15.16}$$

The last relation helps to design the corresponding block diagram (Fig. 15.15).

Assuming $s = j\omega = 0$, the stationary relationship for the constant components is derived

$$\begin{cases} \lambda_0 = \dfrac{b}{b_0}z_0 \\ \delta_0 = \dfrac{a}{a_0}z_0 \end{cases} \tag{15.17}$$

Excluding the variable z

$$\lambda(s) = \frac{-b}{s - b_0}\frac{s + a_0}{a}\delta(s),$$

from which we derive the relationship between the oscillation amplitude, frequency ω and the phase shift ψ as

$$B = \frac{b}{a} \left| \frac{j\omega + a_0}{j\omega - b_0} \right| A,$$

$$\psi = \text{Arg} \left(\frac{a}{s + a_0} \right)_{s=j\omega} - \text{Arg} \left(\frac{-b}{s - b_0} \right)_{s=j\omega} = \text{arctg} \left(\frac{\omega(a_0 + b_0)}{\omega^2 - a_0 b_0} \right).$$

Let us consider the transformation of the harmonic signal as it passes the nonlinearity or modulator as follows:

$$z = \lambda \delta = (\delta_0 + \delta_\sim)(\lambda_0 + \lambda_\sim) = \delta_0 \lambda_0 + \delta_\sim \lambda_0 + \delta_0 \lambda_\sim + \delta_\sim \lambda_\sim = z_0 + z_\sim$$

or

$$\lambda \delta = (\delta_0 + A \sin \omega t)(\lambda_0 + B \sin(\omega t + \psi))$$
$$= \delta_0 \lambda_0 + \delta_0 B \sin(\omega t + \psi) + \lambda_0 A \sin \omega t + AB \sin \omega t \sin(\omega t + \psi).$$

The last term is transformed as

$$AB \sin \omega t \sin(\omega t + \psi) = \frac{1}{2} AB(\cos \psi - \cos(2\omega t + \psi)).$$

The last term is the double frequency harmonic. We know that $|W(j2\omega)| < |W(j\omega)|$, and it legitimates neglecting this term for the approximate analysis. In the case the expression for the variable component balance becomes

$$z_\sim = \delta_\sim \lambda_0 + \delta_0 \lambda_\sim$$

or taking into account expression (15.16)

$$z_\sim = \lambda_0 z_\sim \frac{a}{s + a_0} + \delta_0 z_\sim \frac{-b}{s - b_0},$$

from what we derive the equation

$$b_0[s^2 + (a_0 - b_0)s - a_0 b_0] = [(a_0 - b_0)bs - 2bb_0 a_0]\delta_0.$$

Next, substituting here $s = j\omega$ and analyzing imaginary parts and the real parts separately we arrive at

$$\begin{cases} -\omega^2 b_0 - a_0 b_0^2 = -2ba_0 b_0 \delta_0 \\ (a_0 - b_0)b_0 \omega = (a_0 - b_0)b\delta_0 \omega. \end{cases}$$

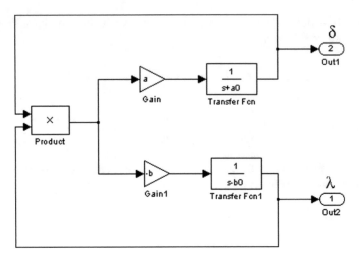

Fig. 15.16 Block diagram of numerical analysis

The system yields the constant component of the consumption fund in GDP δ_0, and the oscillation frequency ω

$$\delta_0 = \frac{b_0}{b},$$

$$\omega = \sqrt{a_0 b_0}.$$

λ_0 is obtained from equation system (15.17)

$$\lambda_0 = \frac{a_0}{a}.$$

Hence, the harmonic linearization defined the oscillation parameters similar to those obtained by Taylor linearization near the equilibrium position.

Numerical analysis of Goodwin model.[4] Figure 15.16 presents the block diagram for Simulink software experiment.[5] Let us obtain the phase portrait in (δ, λ) plane and also the time variations of the variables δ, λ for certain dimensionless parameters and initial conditions.

Example. (a) According to the economic data[6] for Russian Federation, we take the averages of parameters a, a_0, b, b_0 that lead to the following system:

---

[4]The calculation and result analysis are performed by E. M. Fedorova.
[5]It is interesting that direct reduction of the block diagram of Fig. 15.15 by entering gain coefficients to the transfer function numerator leads to incorrect results; that seems to be a Simulink bug.
[6]http://stat.hse.ru.

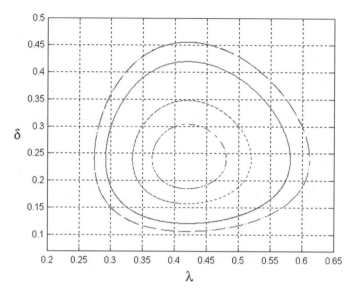

Fig. 15.17 Example. Phase paths obtained close to equilibrium point

$$\begin{cases} \frac{d\delta}{dt} = (2.2\lambda - 0.924)\delta \\ \frac{d\lambda}{dt} = (-1.2\delta + 0.228)\lambda \end{cases} \tag{15.18}$$

According to above-introduced analysis, the center coordinates λ^* and δ^* and also the small oscillation frequencies ω are

$$\lambda^* = \frac{a_0}{a} = 0.42,$$

$$\delta^* = \frac{b_0}{b} = 0.24,$$

$$\omega \approx 0.0731/\text{year},$$

that means that the period of market cycles is to be $T = 1/\omega \approx 13.7$ years.

It should be noted the application of Goodwin model into economy is a formal operation because the economical statistics data does not absolutely correspond to the assumptions are taken as a basis for the model.

The numerical experiment under the fixed initial condition $\delta_0 = 0.3$ and several ones, with respect to λ, $\lambda_0 = [0.3, 0.4, 0.5, 0.6]$, results in the phase portrait and the oscillation phase profiles as shown in Figs. 15.17 and 15.18, correspondingly.

It is evident from the plots that the calculated and experimental results coincide near the equilibrium point at the low oscillation amplitude. The period grows and the oscillations shape deviates from the sinusoidal curve as the amplitude increases.

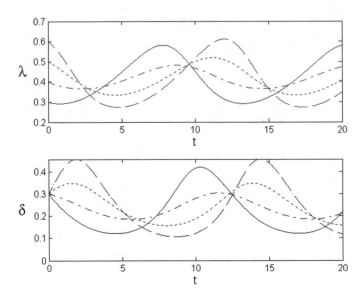

Fig. 15.18   Example. Oscillation histories obtained close to equilibrium point

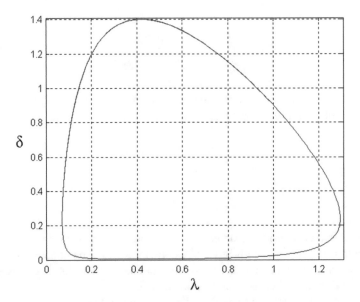

Fig. 15.19   Phase path far away from equilibrium point

Figures 15.19 and 15.20 depict the system oscillations far away from the equilibrium point at the initial conditions $\lambda_0 = 1$, $\delta_0 = 0.9$.

The curves demonstrate that the solution of system (15.18) far away from the equilibrium position is periodic rather than harmonic; the higher the oscillation

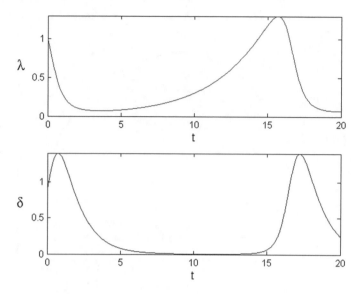

Fig. 15.20 Solutions that are far away from equilibrium

amplitude, the greater is the shape distortion. The solution, generally speaking, has no economic meaning because the initial conditions and the derived oscillation amplitudes are practically unrealistic. The employment share in the total population comes to one only if all the population is economically active. The share of the consumption fund cannot be so high since the consumption is limited by the production. However, the model (15.12) and its analysis can be required in other fields, therefore, we are not going to restrict our analysis only to economically sensible solutions.

Let us develop Goodwin model for the macroeconomic dynamics of Australia. The following system parameter values have been obtained in consideration of regular economic statistics 1978–2003:

$$a = 1.42, b = 0.05, a_0 = 0.58, b_0 = 0.05.$$

Figure 15.21 depicts the phase portrait of system (15.12) plotted based on the above values.

The result is also nonrealistic since the employment in the total population and the consumption fund share cannot exceed 1. Nonetheless limitations were not imposed on the variables in making the model.

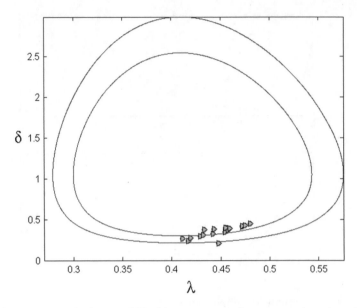

Fig. 15.21 Phase portrait of system 5.52 with system economic parameters from Australia

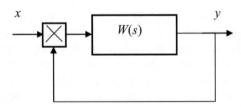

Fig. 15.22 Block diagram of multiplying feedback system

15.4   Frequency Analysis of Product Unit

The performed analysis of the original equations of Goodwin model by their reduction to a block diagram requires to answer more general question on the frequency
characteristics of a single product unit as the element of feedback circuit of the
block diagram in Fig. 15.15. It has been already noted in the previous paragraph that
many economic, biological or engineering systems with bilinear nonlinearity can be
modelled this way.

   Let us consider the harmonic signal as it moves from the input x to the output y
through the system given in Fig. 15.22.

   Although the architecture of the feedback system seems traditional for the classical
control theory, the product element is used instead of summation or a comparator. We
are not able to derive the "close-loop" transfer function for this system. Moreover,
the existence of the product for classifying the system as a linear one for which there

is no transfer function concept at all. We have to consider real and imaginary parts of the harmonic gain; the latter will be the function of frequency, amplitude, and initial conditions.

On the other hand, the scheme in Fig. 15.22 can be considered as an ordinary closed system with a feedback nonstationary element or modulator in the case of the input harmonic x. Then, it is difficult to say about the input harmonic gain factor from x to y: it is either infinite (i.e., a parametric resonance has been excited in the system or the system is not stable) or zero (i.e., a parametric resonance has not been excited). There are only the two conditions for a signal to be transferred from input to output: the system is at the boundary of either parametric resonance excitation or stability. But then, the oscillations are known to have the amplitude and, therefore, the gain factor is governed by the initial conditions. Moreover, it is necessary to remember that transfer function search forces to consider only the second parametric resonance oscillations because just in this case the parameter variation frequency is identical with that of the system oscillations. This means the output oscillations are to bear a constant component.

Let us consider the simplest model for the linear stationary block (see Fig. 15.22)

$$W(s) = \frac{a}{s + a_0}. \tag{15.19}$$

The input signal is to include a constant component equal to a_0/a. The output oscillations vanish if constant component is less than a_0/a; they are divergent when a_0/a exceeds the constant component. However, according to the accepted terminology (see Chap. 2) these oscillations are not the parametric resonance. Indeed, in the case under consideration, the parameter variation make the system sometimes stable, sometimes unstable. Obviously, the oscillation frequency is the frequency of parameter variation. Thus, the stationarization and the search for the excitation conditions of the second parametric resonance are not correct. It is interesting that nevertheless performed analysis yields qualitatively correct necessary value of the constant component and incorrect conclusion on oscillation frequencies.

Fortunately, the original equation with the transfer function of (15.19) type can be integrated in state space easily. This provides the direct solution and the assessment of an input harmonic transfer constant by expanding the latter into series. Indeed, the scheme in Fig. 15.22 with transfer function (15.19) obeys the first-order nonautonomous differential equation in the state space as

$$\dot{y} + a_0 y = ayx,$$

where the initial conditions are

$$y(0) = y_0.$$

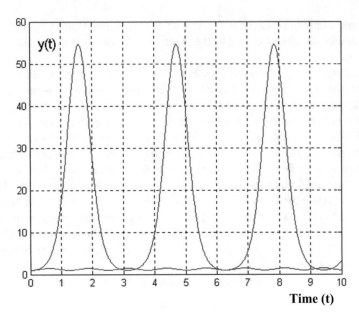

Fig. 15.23 Exact solution of Eq. (15.20) with $a = 1$ and $y(0) = 1$. Upper plot corresponds to $A/\Omega = 2$, lower plot corresponds to $A/\Omega = 0.25$

If the input is a harmonic signal of frequency Ω, amplitude A and the constant component a_0/a we have

$$\dot{y} + a_0 y = ay(a_0/a + A \sin \Omega t)$$

or

$$\dot{y} = ayA \sin \Omega t. \tag{15.20}$$

The variables are separated in Eq. (15.20) which results in the analytical solution

$$y(t) = y_0 e^{\frac{A}{\Omega}(1-\cos \Omega t)}. \tag{15.21}$$

As it can be seen, the exact solution has the constant component and the fundamental frequency oscillations at the parameter variation frequency Ω. Figure 15.23 plots the derived function in respect of small and large values of A/Ω.

It is interesting to observe that, anyway, the behavior of solution (15.21) and original Goodwin equations are similar with the only difference that the large oscillation waves are asymmetric.

To analyze the frequency characteristics of the selected unit, we need to define the zero *th* and the first Fourier expansion terms in the obtained solution

$$y(t) = \frac{b_0}{2} + b_1 \cos \Omega t + b_2 \cos 2\Omega t + \cdots + c_1 \sin \Omega t + c_2 \sin 2\Omega t + \cdots ,$$

where

$$b_k = \frac{2}{T} \int_0^T y(t) \cos k\Omega t \, dt ,$$

$$c_k = \frac{2}{T} \int_0^T y(t) \sin k\Omega t \, dt , \quad k = 0, 1, \ldots .$$

Since the function $y(t)$ is even, $c_k = 0$ and

$$b_k = \frac{4}{T} \int_0^{T/2} y(t) \cos k\frac{2\pi t}{T} \, dt = \frac{2\Omega}{\pi} \int_0^{\pi/\Omega} y(t) \cos k\Omega t \, dt , \quad k = 0, 1, \ldots .$$

Therefore,

$$b_0 = y_0 \frac{2\Omega}{\pi} \int_0^{\pi/\Omega} e^{A/\Omega(1-\cos \Omega t)} dt \tag{15.22}$$

$$b_1 = y_0 \frac{2\Omega}{\pi} \int_0^{\pi/\Omega} e^{A/\Omega(1-\cos \Omega t)} \cos \Omega t \, dt . \tag{15.23}$$

Zero term (15.22) is obtained by deriving the integrals as follows:

$$b_0 = \frac{2\Omega}{\pi} \int_0^{\pi/\Omega} y_0 e^{A/\Omega(1-\cos \Omega t)} dt = \frac{2\Omega}{\pi} e^{A/\Omega} \int_0^{\pi/\Omega} e^{-A/\Omega \cos \Omega t} dt$$

$$= y_0 \frac{2\Omega}{\pi} e^{A/\Omega} \int_0^{\pi/\Omega} \left[1 - A/\Omega \cos \Omega t + \frac{1}{2!} (A/\Omega \cos \Omega t)^2 \right.$$

$$\left. - \frac{1}{3!} (A/\Omega \cos \Omega t)^3 + \ldots \right] dt$$

$$= y_0 \frac{2\Omega}{\pi} e^{A/\Omega} \left[\int_0^{\pi/\Omega} dt - A/\Omega \int_0^{\pi/\Omega} \cos \Omega t dt + \frac{1}{2!} (A/\Omega)^2 \right.$$

$$\int_0^{\pi/\Omega} \cos^2 \Omega t dt - \frac{1}{3!} (A/\Omega)^3 \int_0^{\pi/\Omega} \cos^3 \Omega t dt + \ldots \right]$$

$$= y_0 \frac{2\Omega}{\pi} e^{A/\Omega} \left[\frac{\pi}{\Omega} + \frac{1}{2!} (A/\Omega)^2 \frac{1}{2} \frac{\pi}{\Omega} + \frac{1}{4!} (A/\Omega)^4 \frac{3}{8} \frac{\pi}{\Omega} + \frac{1}{6!} (A/\Omega)^6 \frac{5}{6} \frac{3}{8} \frac{\pi}{\Omega} + \ldots \right]$$

$$= y_0 2 e^{A/\Omega} \left[1 + \frac{1}{2!} (A/\Omega)^2 \frac{1}{2} + \frac{1}{4!} (A/\Omega)^4 \frac{3}{8} + \frac{1}{6!} (A/\Omega)^6 \frac{5}{6} \frac{3}{8} + \ldots \right].$$

The last series can be taken as $y_0 2\, e^{A/\Omega}$ at $A/\Omega < 1$. Otherwise, since

$$b_n < y_0 2 e^{A/\Omega} \left[\frac{1}{n!} (A/\Omega)^n \right], n = 1, 2 \ldots .$$

the following upper estimate is true:

$$b_0 < y_0 2 e^{A/\Omega} \operatorname{ch} A/\Omega = y_0 (e^{2A/\Omega} - 1) \approx y_0 e^{2A/\Omega}.$$

The second integral derivation in (15.23) leads to

$$b_1 = \frac{2\Omega}{\pi} \int_0^{\pi/\Omega} y_0 e^{A/\Omega(1-\cos \Omega t)} \cos \Omega t dt = \cdots =$$

$$= -y_0 2 A/\Omega e^{A/\Omega} \left[\frac{1}{2} + \frac{1}{3!} (A/\Omega)^2 \frac{3}{8} + \frac{1}{5!} (A/\Omega)^4 \frac{5}{6} \frac{3}{8} + \cdots \right].$$

Using the same line of reasoning, as long as the ratio A/Ω is low, the estimate is found as

$$b_1 \approx -y_0 A/\Omega\, e^{A/\Omega}$$

and for large A/Ω as

$$|b_1| < y_0 A/\Omega\, e^{A/\Omega} \mathrm{ch}\, A/\Omega = y_0 A/2\Omega (e^{2A/\Omega} - 1) \approx y_0 A/2\Omega e^{2A/\Omega}.$$

So the transfer coefficients with respect to constant and variable components have been found. The further investigations can be aimed at the system illustrated in Fig. 15.22 with a general transfer function $W(s)$ and also at the application of performed linearization into the analysis of the initial Goodwin combined equations.

Part VI
Multifrequency Oscillations, Stability and Robustness

Annex. The part discusses possible corrections to one-frequency analysis, which accounts certain higher harmonic components. Robustness of wide classes of time-variant and nonlinear systems is evaluated in one-frequency approach. The results on nonlinear and time-variant robustness develop the author's previous findings [11, 15–18] and, finally, the thesis work [19].

Chapter 16
Single-Frequency Approximation Correction of Periodically Nonstationary Systems

The direct solution of the problem of multifrequency process analysis meets enormous difficulties. Indeed, the simplest calculation, for example, of a two-frequency process in a dynamic single-frequency parameter system has three frequencies, three amplitudes, and three phases as unknown quantities. Finding a solution in the nine-dimensional space of unknowns concerning a system with many state coordinates is not a simple problem even using modern computer power.

As usual, the ability to predict process behavior reduces the complexity substantially.

16.1 Basic Equation

Let us consider an autonomous T-periodic system given in the following operator form:

$$G(s) x(t) + H(s) y(t) = 0, \tag{16.1}$$

$$y(t) = a(t - \tau)x(t), \tag{16.2}$$

$$W(s) = H(s)/G(s), \tag{16.3}$$

where $a(t)$ is an arbitrary T-periodic parameter.

Suppose a linear stationary subsystem has the transfer function $W(s)$ and the resonance frequencies ω_1, ω_2, ω_3. According to those frequencies, the system can have a motion composed of three components

$$x(t) = x_1(t) + x_2(t) + x_3(t). \tag{16.4}$$

Output signal (16.2) from the parameter $a(t)$

$$y(t) = a(t-\tau)[x_1(t) + x_2(t) + x_3(t)] \tag{16.5}$$

© Springer International Publishing AG, part of Springer Nature 2017
L. Chechurin and S. Chechurin, *Physical Fundamentals of Oscillations*,
https://doi.org/10.1007/978-3-319-75154-2_16

has components of many frequencies. The mth output harmonic includes components of the nth input harmonic. The coupling coefficients of the harmonics can be evaluated in the form

$$a_{mn} = \frac{2j}{T_m} \int_0^{T_m} a(t - \tau)x_n(t)e^{j\omega_m t} dt,$$

$$T_m = 2\pi/\omega_m, \quad m, n = 1, 2, 3 \ldots. \tag{16.6}$$

The above assumption of the existence of only three basic frequencies enables to select the three components at the periodic element output:

$$y(t) = y_1(t) + y_2(t) + y_3(t). \tag{16.7}$$

Components in Eq. (16.7) can be expressed in terms of the input components using coupling coefficients in Eq. 16.6) as

$$y_1(t) = a_{11}x_1(t) + a_{12}x_2(t) + a_{13}x_3(t)$$
$$y_2(t) = a_{21}x_1(t) + a_{22}x_2(t) + a_{23}x_3(t) \tag{16.8}$$
$$y_3(t) = a_{31}x_1(t) + a_{32}x_2(t) + a_{33}x_3(t).$$

Taking into account expressions (16.4) and (16.7), the system description (16.1) takes the form

$$G(s)[x_1(t) + x_2(t) + x_3(t)] + H(s)[y_1(t) + y_2(t) + y_3(t)] = 0. \tag{16.9}$$

Transferring from description (16.9) to its frequency representation at $s = j\omega$ and grouping the result terms by their frequencies, the following linear equations can be obtained:

$$G(j\omega_1) + H(j\omega_1)[a_{11}x_1(t) + a_{12}x_2(t) + a_{13}x_3(t)] = 0$$
$$G(j\omega_2) + H(j\omega_2)[a_{21}x_1(t) + a_{22}x_2(t) + a_{23}x_3(t)] = 0 \tag{16.10}$$
$$G(j\omega_3) + H(j\omega_3)[a_{31}x_1(t) + a_{32}x_2(t) + a_{33}x_3(t)] = 0.$$

Linear system of Eq. (16.10) yields a solution when its determinant equals to zero, i.e.,

$$\begin{vmatrix} 1 + a_{11}W(j\omega_1) & a_{12}W(j\omega_1) & a_{13}W(j\omega_1) \\ a_{21}W(j\omega_2) & 1 + a_{22}W(j\omega_2) & a_{23}W(j\omega_2) \\ a_{31}W(j\omega_3) & a_{32}W(j\omega_3) & 1 + a_{33}W(j\omega_3) \end{vmatrix} = 0. \tag{16.11}$$

Determinant (16.11) is the core of the known infinite-order Hill determinant and Eq. (16.11) is the approximate multifrequency parametric resonance excitation condition.

Let a single-frequency parameter be taken as

$$a(t - \tau) = a \sin \Omega(t - \tau) = \frac{a}{2j}[e^{+j(\Omega t - \varphi)} - e^{-j(\Omega t - \varphi)}], \quad \varphi = \Omega \tau. \quad (16.12)$$

In that case, the three natural frequencies ω_1, ω_2, and ω_3 can be denoted as $\Omega - \omega$, ω, and $\Omega + \omega$ and two coupling coefficients a_{31} and a_{13} are equal to zero. The coefficients a_{11}, a_{22}, and a_{33} reflect the modulation when the output and input frequencies are the same. Among the frequencies satisfying the relation $\omega = l\Omega/2$, where l is an arbitrary integer, the frequency coincidence $\Omega \pm \omega = \omega$ can be given in the form

$$\frac{2\omega}{l} \pm \omega = \omega \quad (16.13)$$

and it is satisfied only if $l = 1$. This case of multifrequency oscillations including the first parametric resonance will be considered later in Sect. 2.1. And now excluding the first parametric resonance, we assume that

$$a_{11} = a_{22} = a_{33} = 0. \quad (16.14)$$

Thus, the multifrequency parametric excitation condition with respect to a single-frequency parameter variation takes the form

$$\begin{vmatrix} 1 & a_{12}W(j\omega_1) & 0 \\ a_{21}W(j\omega_2) & 1 & a_{23}W(j\omega_2) \\ 0 & a_{32}W(j\omega_3) & 1 \end{vmatrix} = 0$$

or

$$1 - a_{12}a_{21}W(j\omega_1)W(j\omega_2) - a_{23}a_{32}W(j\omega_2)W(j\omega_3) = 0. \quad (16.15)$$

A two-frequency process is going to be considered further because it is widely used in practice. The system is assumed to have two natural frequencies, ω_1 and ω_2. There, we distinguish two main cases depending on the parameter variation frequency Ω: (1) *difference parametric resonance* when the frequency relation

$$\omega_2 - \omega_1 = \Omega \quad (16.16)$$

takes place and (2) *sum parametric resonance* if

$$\omega_2 + \omega_1 = \Omega. \quad (16.17)$$

In both cases, excitation conditions follow from condition (16.15) and have the general form

$$1 - a_{12}a_{21}W(j\omega_1)W(j\omega_2) = 0, \tag{16.18}$$

which however differs for the specific cases of difference and sum resonances.

16.2 Difference Parametric Resonance

The difference parametric resonance also has several insignificant cases. The first case is realized when

$$\omega_1 = \omega_0, \omega_2 = \omega_0 + \Omega \tag{16.19}$$

and $\Omega < \omega_0$. Condition (16.18) defines now the ultraharmonic oscillation excitation relative to the parameter variation frequency Ω. If $\Omega > \omega_0$, the excited oscillations have ultraharmonic and subharmonic components.

Eventually, if the frequency relations

$$\omega_1 = \omega_0 - \Omega, \omega_2 = \omega_0, \tag{16.20}$$

hold, ultraharmonic oscillations and subharmonic oscillations can exist in the ranges $\Omega < \omega_0/2$ and $\omega_0 > \Omega > \omega_0/2$, correspondingly. Excitation condition (16.18) is the same for all the cases and is found from harmonic coupling coefficients. Let us derive the latter concerning case (16.20) according to expressions (16.6) and (16.12) as follows:

$$a_{12} = \frac{2j}{T_1} \int_0^{T_1} a \sin \Omega(t - \tau) \sin \omega_0 t \, e^{-j(\omega_0 - \Omega)t} dt = -\frac{a}{2j} e^{+j\varphi} \tag{16.21}$$

$$a_{21} = \frac{2j}{T_2} \int_0^{T_2} a \sin \Omega(t - \tau) \sin(\omega_0 - \Omega)t \, e^{-j\omega_0 t} dt = +\frac{a}{2j} e^{-j\varphi}. \tag{16.22}$$

The excitation condition arises from condition (16.18):

$$1 - \frac{a^2}{4} W[j(\omega_0 - \Omega)]W(j\omega_0) = 0. \tag{16.23}$$

This means the vectors $W[j(\omega_0 - \Omega)]$ and $W(j\omega_0)$ in the frequency plane have opposite phases and satisfy the modulus equality

Fig. 16.1 Difference
parametric resonance

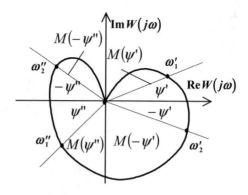

Fig. 16.2 Series parametric
resonance system

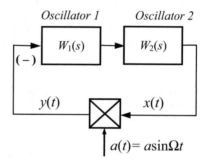

$$\left| \frac{1}{W(j\psi)} \right| = \frac{a^2}{4} |W(-j\psi)|, \qquad (16.24)$$

as illustrated in Fig. 16.1 in the Nyquist plane as an example, where $M(\psi)$ is the frequency response modulus.

Example. The difference parametric resonance is sometimes called a series parametric resonance because it happens in a system consisting of two consequentially connected oscillatory circuits (it is not to be mixed up with a series resonance in an RLC electric circuit).

Let us take the tandem of oscillatory circuits

$$W(s) = W_1(s)W_2(s) = \left[\frac{1}{s^2 + 0.1s + 1} \right] \left[\frac{1}{0.01s^2 + 0.001s + 1} \right],$$

as given in Fig. 16.2.

The frequency characteristic $W(j\omega)$ presented in Fig. 16.3 has resonance picks at $\omega_1 = 1$ rad/s and $\omega_2 = 10$ rad/s. Figure 16.4 shows an excitation process in the difference parametric resonance system. Table 16.1 lists the comparative results of approximate critical parameter evaluation and a computer simulation.

Fig. 16.3 Example.
Frequency response of the
system

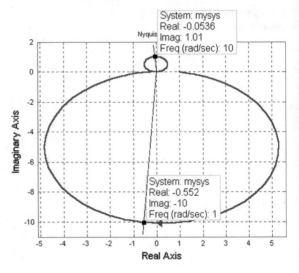

Fig. 16.4 Example.
Difference parametric
resonance excitation process

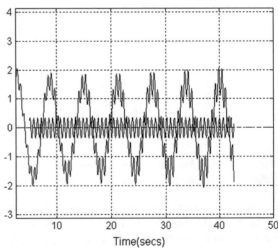

Table 16.1 Example. Excitation boundaries of difference parametric resonance (analytical and experiment)

Ω	8.5	8.8	8.9	9	9.1	9.2	9.5
a_{an}	2.22	1.34	0.96	0.63	0.9	1.21	1.73
$a_{\exp}$	3.0	1.3	0.85	0.6	0.85	1.4	3.8

16.3 Sum Parametric Resonance

Due to (16.17), the sum parametric resonance oscillations contain two natural frequencies

$$\omega_1 = \omega_0, \, \omega_2 = \Omega - \omega_0, \, \Omega > \omega_0. \tag{16.25}$$

As both natural frequencies are lower than Ω, low-frequency process containing two subharmonic components can be observed. Coupling coefficients (16.6) take the forms

$$a_{12} = \frac{2j}{T_1} \int_0^{T_1} a \sin \Omega(t - \tau) \sin(\Omega - \omega_0)t \, e^{-j\omega_0 t} dt = -\frac{a}{2j} e^{-j\varphi} \tag{16.26}$$

$$a_{21} = \frac{2j}{T_2} \int_0^{T_2} a \sin \Omega(t - \tau) \sin \omega_0 t \, e^{-j(\Omega - \omega_0)t} dt = -\frac{a}{2j} e^{-j\varphi} \tag{16.27}$$

and excitation condition (16.18) is written in the form

$$1 + \frac{a^2}{4} e^{-j2\varphi} W(j\omega_0) W[j(\Omega - \omega_0)] = 0. \tag{16.28}$$

From this, we see that the sum parametric resonance excitation condition covers previous condition (16.23) at $\varphi = \pi/2$. Figure 16.5 depicts conditions (16.28) on the Nyquist plane.

The sum parametric resonance is excited if the resonance pick $W(j\omega_0)$ is out of the circle of radius $r = 4a^{-2} |W[j(\Omega - \omega_0)]|^{-1}$. An illustration like that is also possible on the inverse Nyquist hodograph plane.

The assumption of only two resonance picks existence is to be remembered every time in the applications of excitation condition (16.18). Moreover, it should be noted that the sum parametric resonance appears in case if at least one of two vectors $W(j\omega_0)$, $W(j\Omega - j\omega_0)$ exceeds $0.5a$ in modulus. On the other hand, the excitation does

Fig. 16.5 Sum parametric resonance excitation condition

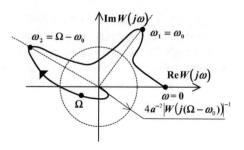

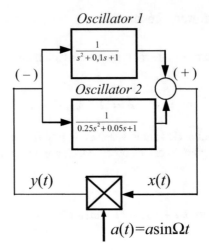

Fig. 16.6 Example. Parallel parametric resonance system

Table 16.2 Example. Sum parametric resonance excitation boundaries (analytical and experiment)

Ω	2.5	2.8	2.9	3.0	3.1	3.2	3.5
a_{an}	0.52	0.3	0.18	0.20	0.23	0.3	0.44
a_{exp}	0.7	0.3	0.23	0.23	0.27	0.4	0.57

not exist if either of the vectors is less than $0.5a$ in modulus. The same requirements to the vectors $W(j\omega_0)$ and $W(j\Omega - j\omega_0)$ can be formulated for the difference parametric resonance.

Example. The sum parametric resonance is often called a parallel parametric resonance since it takes place in a dynamic system comprising two in-parallel resonance circuits. Figure 16.6 displays an example of a similar system.

The stationary subsystem has the transfer function

$$W(s) = W_1(s) + W_2(s) = \frac{1}{s^2 + 0.1s + 1} + \frac{1}{0.25s^2 + 0.05s + 1}$$

and the frequency response is given in Fig. 16.7.

The characteristic bears two resonance picks at the frequencies $\omega_1 = 1$ rad/s and $\omega_2 = 2$ rad/s. The sum parametric resonance can be observed, for example, at the parameter variation frequency $\Omega = 3$ rad/s (see Fig. 16.8). Table 16.2 compares the analytical evaluation and simulation results.

Fig. 16.7 Example. System
frequency response

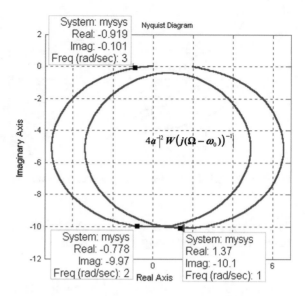

Fig. 16.8 Example. Sum
parametric resonance
excitation process

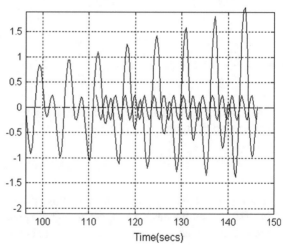

16.4   Constant Component Parametric Resonance Excitation

It was already pointed out in Chap. 2 that the second parametric resonance can be
considered as a special case of combined (two-frequency) resonance in which one
of the frequencies is zero. A constant process component appears in a system along
with the parametric resonance excitation if one of the frequencies of parametric
oscillations and that of parameter variations are the same. The oscillations include
constant (x_0) and variable $(\tilde{x}(t) = \sin \Omega t)$ components during both the sum and

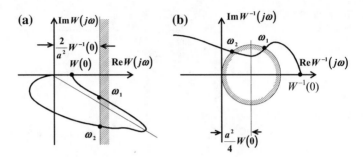

Fig. 16.9 Combined parametric resonance excitation condition

difference resonance cases with $\omega_0 = \Omega$. The coupling coefficients of harmonics have to be derived separately although as follows:

$$a_{12} = \frac{1}{T}\int_0^T a(t - \tau)\tilde{x}(t)\mathrm{d}t = \frac{1}{T}\int_0^T a\,\sin(\Omega t - \Omega \tau)\sin \Omega t \mathrm{d}t = \frac{a}{4}(e^{+j\varphi} + e^{-j\varphi})$$

$$\tag{16.29}$$

$$a_{21} = \frac{2j}{T}\int_0^T a(t - \tau)e^{-j\Omega t}\mathrm{d}t = \frac{2j}{T}\int_0^T a\,\sin(\Omega t - \Omega \tau)e^{-j\Omega t}\mathrm{d}t = ae^{-j\varphi}. \tag{16.30}$$

Substituting coefficients (16.29) and (16.30) in (16.18), the excitation condition is obtained in the form

$$1 - \frac{a^2}{4}(1 + e^{-j2\varphi})W(0)W(j\Omega) = 0. \tag{16.31}$$

Condition (16.31) has its graphical illustration on Nyquist hodograph plane (Fig. 16.9a) or on the inverse Nyquist hodograph plane (Fig. 16.9b). The parametric resonance is excited as long as the point $\omega = \Omega$ belongs to the shaded area.

Example. A parametrically excited oscillator has the transfer function of its linear part as

$$W(p) = \frac{1}{s^2 + 0.2s + 1}$$

and its natural frequency is about $\omega_0 = 1$ rad/s. Figure 16.10 presents the simulation results for the parametric excitation under the feedback parameter variation frequency $\Omega = 0.9$ rad/s. The asymmetrical process with the constant component takes place. The analytical and simulated results are compared in Table 16.3.

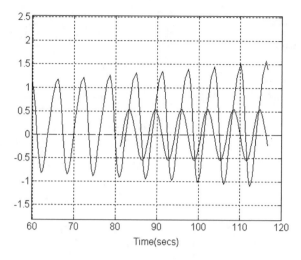

Fig. 16.10 Example. Second parametric resonance excitation process

Table 16.3 Example. Combined parametric resonance excitation boundaries (analytical and experiment)

Ω	0.75	0.85	0.9	0.95	0.99
a_{an}	0.98	0.88	0.85	0.96	1.98
a_{exp}	1.35	1.0	0.9	0.9	1.0

16.5 First Parametric Resonance at Two-Frequency Approximation

As follows from expression (16.13), the first parametric resonance can be considered at two-frequency approximation. Let us assume that $l = 1$ in (16.13) and there are two oscillation process frequencies

$$\omega_1 = \Omega/2, \omega_2 = \omega_1 + \Omega = \frac{3}{2}\Omega. \tag{16.32}$$

Since the frequency $\omega_3 = 0$,

$$a_{13} = a_{32} = a_{23} = 0$$

and $a_{22} = 0$ owing to a single-frequency variation of parameter (16.12). As a result, solution (16.11) takes on the form

$$\begin{vmatrix} 1 + a_{11}W(j\frac{\Omega}{2}) & a_{12}W(j\frac{\Omega}{2}) \\ a_{21}W(j\frac{\Omega}{2}) & 1 \end{vmatrix} = 0$$

Fig. 16.11 Structural representation of corrected parametric resonance

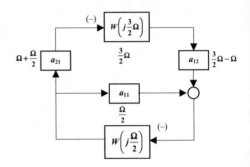

or

$$1 + W\left(j\frac{\Omega}{2}\right)\left[a_{11} - a_{12}a_{21}W\left(j\frac{3}{2}\Omega\right)\right] = 0. \tag{16.33}$$

The structural representation in Fig. 16.11 has circuits for both the first and difference parametric resonances. The coupling coefficient a_{11} is the same circle (4.13) of the first parametric resonance as before, i.e.,

$$a_{11} = W(j\varphi) = -\frac{a}{2j}e^{-j\varphi}, \varphi = \Omega\tau, \Omega = \frac{2\pi}{T} \tag{16.34}$$

where the constant component of parameter variation is attributed to the linear stationary part $W(s)$, and a_0 is assumed to be zero. The rest of the coefficients agree with ones (16.21) and (16.22) for the difference parametric resonance, i.e.,

$$a_{12} = \frac{2j}{T}\int_0^T a\sin\Omega(t-\tau)\sin\frac{3}{2}\Omega t\,e^{-j\frac{\Omega}{2}t}dt = -\frac{a}{2j}e^{+j\varphi} \tag{16.35}$$

$$a_{21} = \frac{2j}{T}\int_0^T a\sin\Omega(t-\tau)\sin\frac{\Omega}{2}t\,e^{-j\frac{3}{2}\Omega t}dt = +\frac{a}{2j}e^{-j\varphi}. \tag{16.36}$$

Thus, the excitation condition of the first parametric resonance follows from expressions (16.33)–(16.36) in the corrected form as

$$1 + W\left(j\frac{\Omega}{2}\right)W(j\varphi) = 1 + W\left(j\frac{\Omega}{2}\right)\left[-\frac{a^2}{4}W\left(j\frac{3}{2}\Omega\right) - \frac{a}{2j}e^{-j\varphi}\right] = 0. \tag{16.37}$$

The correction in the square brackets indicates the central circle shift of $0.25a^2W(j3\Omega/2)$, as displayed in Fig. 16.12. It is obvious that the opposite phase

Fig. 16.12 Frequency correction to first parametric resonance excitation condition

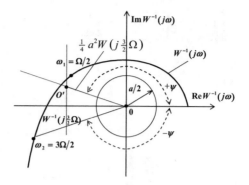

Fig. 16.13 Example. Two-frequency excitation condition calculation

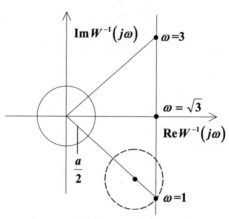

arrangement of the vectors $W(j\Omega/2)$ and $W(j3\Omega/2)$ is preferable to excite the parametric resonance.

With $a_0 \neq 0$, the parameter transfer function corrected subject to the third harmonic is

$$W(j\varphi) = a_0 - \frac{a^2}{4}W\left(j\frac{3}{2}\Omega\right) - \frac{a}{2j}e^{-j\varphi}.$$

Example. Let us consider a positive T-periodic feedback dynamic system as an example. A stationary subsystem has the transfer function

$$W(s) = \frac{s}{(0.6s + 1)^2}$$

and the inverse Nyquist hodograph as depicted in Fig. 16.13.

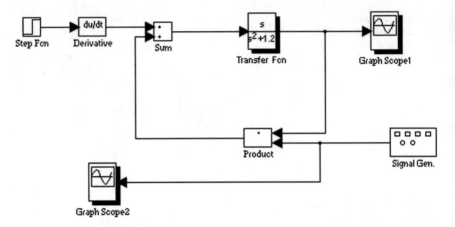

Fig. 16.14 Example. Experiment block diagram

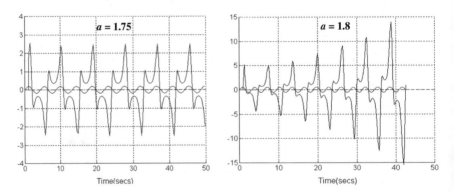

Fig. 16.15 Example. Oscillation excitation processes

As appears from the frequency response plot, two points at the frequencies ω_1 и ω_3 have opposite phases and parametric excitation on the frequency $\Omega = 2$ rad/s leads to the two-frequency parametric resonance. Figures 16.14 and 16.15 illustrate the Simulink model structure and the excitation processes, correspondingly.

16.6   Invariance of First Parametric Resonance Excitation Conditions

Realistic calculations of nonstationary systems result quit often in the coincidence of parametric resonance excitation conditions in the different approximations of the same oscillatory object model. The knowledge of coincidence conditions (or invariance) allows avoiding cumbersome transformations to improve the model, the analysis of system can be done with simpler equations.

Let us consider the first system model described by a difference equation in the operator form

$$\sum_{k=0}^{n} s^k[a_k(t-\tau)x(t)] = \sum_{k=0}^{n} \{A_k s^k x(t) + B_k s^k[f_k(t-\tau)x(t)]\} = 0, \quad (16.38)$$

where the parameters in $a_k(t-\tau) = A_k + B_k f_k(t-\tau)$ alternates as $f_k(t-\tau) = f_k(t-\tau+T)$ with a period $T = 2\pi/\Omega$ and τ is a time shift between oscillations of the parameters and the $x(t)$ coordinate.

We assume the system in focus has the first parametric resonance, that is to say, Eq. (16.38) has an approximate stable solution and its derivatives

$$x(t) = \exp(-j\Omega t/2), \quad x^{(k)}(t) = (-j\Omega/2)^{kx(t)}. \quad (16.39)$$

The known relations between the amplitudes and k derivatives of the first harmonic of a parameter function arise from the periodic parameter representation

$$f_k(t-\tau) = \sum_{l=-\infty}^{l=+\infty} f_{k\ell} \exp[j\Omega l(t-\tau)]$$

in the form

$$f_{k1}^{(k)} = (j\Omega)^k f_{k1}^{(0)} \exp(-j\varphi), \quad \varphi = \Omega\tau \quad (16.40)$$

Taking into account relations (16.39) and (16.40) and using Leibniz formulae and Newton's binomial, the term $p^k[f_k(t-\tau)x(t)]$ in Eq. (16.38) is derived as follows:

$$\sum_{m=0}^{k} C_k^m f_k^{(k-m)}(t-\tau)x^{(m)}(t) \cong e^{-j\varphi} \sum_{m=o}^{k} C_k^m (j\Omega)^{k-m} f_{k1}^{(0)}(-j\Omega/2)^m x(t)$$

$$= e^{-j\varphi} f_{k1}^{(0)} x(t) \sum_{m=o}^{k} C_k^m (j\Omega)^{k-m}(-j\Omega/2)^m = f_{k1}^{(0)} x(t)(j\Omega/2)^k e^{-j\varphi} \quad (16.41)$$

because $\sum_{m=0}^{k} C_k^m 2^{k-m}(-1)^m = 1$. Substituting expressions (16.40) into Eq. (16.38), the approximate frequency equality is obtained as

$$\sum_{k=0}^{n} (A_k + f_{k0} B_k)(j\Omega/2)^k + \sum_{k=0}^{n} B_k f_{k1}(j\Omega/2)^k e^{-j\varphi} = 0$$

which gives the approximate first parametric resonance excitation conditions

$$\left| \sum_{k=0}^{n} (A_k + f_{k0}B_k)(j\Omega/2)^k \right| = \left| \sum_{k=0}^{n} B_k f_{k1}(j\Omega/2)^k \right| \qquad (16.42)$$

as a result obtained by the first model of the system under study.

Let us consider now the second model of a system described by the differential equation

$$\sum_{k=0}^{n} a_k(t - \tau)s^k x(t) = 0. \qquad (16.43)$$

The approximate parametric resonance excitation condition to equation like (16.43) is written down in the form

$$\left| \sum_{k=0}^{n} (A_k + f_{k0}B_k)(j\Omega/2)^k \right| = \left| \sum_{k=0}^{n} B_k f_{k1}(-j\Omega/2)^k \right|. \qquad (16.44)$$

The solutions of Eqs. (16.38) and (16.43) are generally different with regard to stability loss conditions (16.42) and (16.44). At the same time, the conditions agree with each other in the cases when (a) all the f_{k1} numbers are real, (b) all the f_{k1} numbers are imaginary, (c) all the f_{k1} numbers are equal, and (d) some of the f_{k1} numbers are zero. These requirements can be reformulated in terms of periodic functions $f_k(t)$ by the properties of Fourier coefficients: all the functions are even, all the functions are odd symmetric, all the functions are equal; some of the functions can be constants.

The identity of conditions (16.42) and (16.44), when the above formulated requirements held, implies that at least the first parametric resonance excitation regions are the same for all models.

Chapter 17
Single-Frequency Approximation Correction of Nonlinear Systems

This chapter provides the multiple-frequency analysis of the nonlinear dynamic system oscillations. We are interested in the influence of the higher harmonics on the evaluations given in Chap. 3 by the first harmonic approximation.

17.1 Higher Harmonic Components

Higher harmonic gains of nonlinear element. As mentioned in Chap. 3, with a single-frequency input harmonic, Fourier expansion of the output signal of the nonlinear element $F(x)$ bears higher harmonics. Their profile substantially depends on the type of the function $F(x)$. Therefore, the evaluation of higher harmonic components is needed.

Using known Fourier coefficients (7.5) and (7.5), the nonlinear element gains from the amplitude A of the first harmonic to the m th harmonic amplitude are

$$q'_m = \frac{1}{\pi A} \int_0^{2\pi} F(A \sin \psi) \sin m\psi \, d\psi \tag{17.1}$$

$$q''_m = \frac{1}{\pi A} \int_0^{2\pi} F(A \sin \psi) \cos m\psi \, d\psi \tag{17.2}$$

or in the complex representation

$$W_m(A) = q'_m + j q''_m = \frac{j}{\pi A} \int_0^{2\pi} F(A \sin \psi) e^{-jm\psi} \, d\psi. \tag{17.3}$$

© Springer International Publishing AG, part of Springer Nature 2017
L. Chechurin and S. Chechurin, *Physical Fundamentals of Oscillations*,
https://doi.org/10.1007/978-3-319-75154-2_17

Fig. 17.1 Structure
representation for the
analysis of the m th harmonic

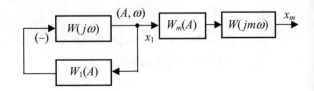

Thus, the m th harmonic x_m is connected with the first harmonic $x_1 = A \sin \psi$ by
the relation

$$x_m = (q'_m + jq''_m)x_1 = W_m(A)x_1. \tag{17.4}$$

Examples. Harmonic linearization coefficient was already evaluated for an ideal two-
position relay, see (7.21). The higher harmonic linearization factors can be evaluated
as follows:

$$W_m(A) = q'_m = \frac{1}{\pi A} \int_0^{2\pi} F(A \sin \psi) \sin m\psi \, d\psi$$

$$= \frac{1}{\pi A} \left[\int_0^\pi B \sin m\psi \, d\psi - \int_\pi^{2\pi} B \sin \psi \, d\psi \right]$$

$$= \frac{B}{m\pi A}(1 + \cos 2m\pi - 2\cos m\pi)$$

or in the notation

$$W_m(A) = \frac{2B}{m\pi A}[1 - (-1)^m].$$

Here, the gain is zero for even subscripts m, and

$$W_m(A) = \frac{4B}{m\pi A} \tag{17.5}$$

otherwise.

To obtain the characteristic of a real two-position relay, the gains are derived in
the similar manner by referring to Fig. 7.3

$$W_m(A) = \frac{j}{\pi A} \int_0^{2\pi} F(A \sin \psi) e^{-jm\psi} d\psi$$

$$= \frac{jB}{\pi A} \left[-\int_0^{\psi_1} e^{-jm\psi} d\psi + \int_{\psi_1}^{\pi+\psi_1} e^{-jm\psi} d\psi - \int_{\pi+\psi_1}^{2\pi} e^{-jm\psi} d\psi \right]$$

$$= \frac{2B}{k\pi A} e^{-jm\psi_1} [1 - (-1)^m],$$

where

$$W_m(A) = \frac{4B}{m\pi A} e^{-jm\psi_1} \qquad (17.6)$$

for the odd subscripts m, and $\psi_1 = \arcsin \frac{b}{A}$ as before.

Higher harmonic component assessment. The previous nonlinear dynamic system description is assumed as

$$G(s)x + H(s)F(x) = 0 \qquad (17.7)$$

Let the forced or autonomous periodic motion of the system contain the first and the m th harmonics

$$x(t) = x_1 + x_m = A \sin \omega t + A_m e^{-j\psi_m} \sin \omega t. \qquad (17.8)$$

Then, nonlinear system description (17.7) takes the form

$$G(s)(x_1 + x_m) + H(s)F(x_1 + x_m) = 0. \qquad (17.9)$$

According to the single-frequency harmonic linearization method (see Chap. 3), the m th harmonic is considered to become small as the signal passes the linear dynamic subsystem, so it is neglected at the nonlinear element input. In this case, instead of (17.9) the approximate description is given. It is based on harmonic linearization of the nonlinear element with respect to the first and the m th harmonics as

$$G(s)(x_1 + x_m) + H(s)[W_1(A) + W_m(A)]x_1 = 0. \qquad (17.10)$$

Equation (17.10) can be divided into the two on the frequency basis:

$$G(j\omega)x_1 + H(j\omega)W_1(A)x_1 = 0, \qquad (17.11)$$

$$G(jm\omega)x_m + H(jm\omega)W_m(A)x_1 = 0. \qquad (17.12)$$

Equation (17.11) is the first harmonic approximation studied in Chap. 3; the m th harmonic appears from (17.12) as

$$x_m = -\frac{H(jm\omega)}{G(jm\omega)} W_m(A)x_1 = -W(jm\omega)W_m(A)x_1, \qquad (17.13)$$

where its modulus and phase are

$$|x_m| = |W(jm\omega)|\,|W_m(A)|\,A \qquad (17.14)$$

$$\psi_m = \arg[-W_m(A)] + \arg W_m(A). \qquad (17.15)$$

The system structure representation depicted in Fig. 17.1 corresponds to Eqs. (17.11) and (17.12). Equations (17.14) and (17.15) enable the assessment of the m th harmonic if the amplitude and frequency of the first harmonic are derived from (17.11). The further correction of the first harmonic approximation is not needed if the order of vanishing of the m th harmonic is equal or higher than the order of the first harmonic. In other words, Eq. (17.11) is accurate enough in this case. Otherwise, either at least the first approximation correction or the complete two-frequency analysis is to be performed.

17.2 Harmonic Linearization Factor Correction

If the higher harmonic evaluation is found to be small but insufficiently small to be neglected in the system, one can evaluate its influence on first harmonic approximation. For that purpose, the transformation of Eq. (17.9) to an improved, but still approximate description differing from definition (17.16) is performed. Let the nonlinear function $F(x)$ in Eq. (17.9) have an increment, i.e.,

$$G(s)(x_1 + x_m) + H(s)[F(x_1) + \Delta F] = 0, \qquad (17.16)$$

here ΔF is the first harmonic correction component due to the higher harmonic existence. The component is expressed in the form

$$\Delta F = \left.\frac{dF(x)}{dx}\right|_{x=x_1} x_m. \qquad (17.17)$$

According to Chap. 5, the first harmonic frequency component is distinguished in (17.17) as

$$\Delta F_1 = a_{1m}x_m, \qquad (17.18)$$

where a_{1m} is the coupling coefficient with respect to the first and m th harmonics. Now, harmonically linearized Eq. (17.16) takes on the form

Fig. 17.2 Block diagram for harmonic linearization factor correction

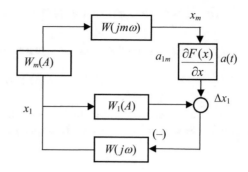

$$G(s)(x_1 + x_m) + H(s)[W_1(A)x_1 + W_m(A)x_1 + a_{1m}x_m] = 0. \qquad (17.19)$$

Grouping the Eq. (17.19) around the frequencies, as before, we arrive at two equations

$$G(j\omega)x_1 + H(j\omega)[W_1(A) + a_{1m}x_m] = 0 \qquad (17.20)$$

$$G(jm\omega)x_m + H(jm\omega)W_m(A)x_1 = 0. \qquad (17.21)$$

Equation (17.21) is the same as Eq. (17.12) derived above. Substituting x_m in Eq. (17.20) for the same parameter from Eq. (17.21), the frequency equation is deduced as

$$G(j\omega)x_1 + H(j\omega)\left[W_1(A) - a_{1m}\frac{H(jm\omega)}{G(jm\omega)}W_m(A)\right]x_1 = 0. \qquad (17.22)$$

Figure 17.2 depicts the block diagram described by the frequency equation above.

As appears from Fig. 17.2, the correction means the introduction of a high-frequency dynamic circuit which is parallel to the first harmonic linearization factor. The most striking difference of the periodic parameter in the correction circuit of a nonlinear system from that of the linear periodically nonstationary systems studied before is the absence of its time shift ($\tau = 0$) with respect to the coordinate. In turn, it means zero phase shift ($\varphi = 0$). This can be explained by rigid synchronization of the nonlinear function derivative with its periodic argument variation.

As a concluding remark, we noted that the practical value of these correction evaluations is not so much in its proper numerical assessment but rather in the possibility to conclude if the first harmonic approximation is enough.

17.3 Harmonic Stationarization Factor Correction

Let us consider again the nonlinear dynamic system

$$G(s)x(t) + H(s)F[x(t)] = H(s)f(t).$$

Let the forced or free $(f = 0)$ oscillations $x_1(t)$ with a frequency ω_1 and an amplitude A exist in the system. It means that the harmonically linearized equation

$$G(j\omega_1)x_1(t) + H(j\omega_1)W_1(A)x_1(t) = H(j\omega_1)f(t) \qquad (17.23)$$

is satisfied. The stability of both forced and free oscillations is known to obey the incremental equation

$$G(p)\Delta x(t) + H(p)\left(\frac{\partial F(x)}{\partial x}\right)_{x=x_1(t)} \Delta x(t) = 0. \qquad (17.24)$$

For simplicity, the output characteristic of the nonlinear element $F(x)$ is assumed to be odd-symmetrical and there are all odd harmonics of frequencies $m\omega_1$, $m = 1$, 3, 5, … in the input harmonic $x_1(t)$. Besides, the derivative of a periodic argument function (periodic parameter) varies at frequency $\Omega = 2\omega_1$ and Fourier expansion of the parameter includes $k\Omega$ harmonic components, $k = 1, 2, …$ with a_k amplitudes. So, all the excitation conditions for combined parametric oscillations provided in Chap. 15 can be applied here to define the stability loss of the oscillations $x_1(t)$.

Let the increment $\Delta x(t)$ in Eq. (17.24) consist of the two frequency components, ω_1 and $n\omega_1$ as

$$\Delta x_1(t) = \Delta x_1 + \Delta x_n \qquad (17.25)$$

or, in other words, let the system have two natural frequencies $\omega_1 = \Omega/2$ and $\omega_2 = n\Omega/2, n = 2, 3, …$ Then the multifrequency parameter $\left(\frac{\partial F(x)}{\partial x}\right)_{x=x_1(t)}$ contains four fundamental frequency components: Ω, $n\Omega$, $(n - 1)\Omega/2$, and $(n + m)/\Omega$.

In the above notations, the frequency separation of Eq. (17.24) gives two equations

$$G(j\omega_1)\Delta x_1 + H(j\omega_1)[a_{11}\Delta x_1 + a_{1n}\Delta x_n] = 0$$

$$QG(j\omega_n)\Delta x_n + H(j\omega_n)[a_{n1}\Delta x_1 + a_{nn}\Delta x_n] = 0. \qquad (17.26)$$

Excluding the increment Δx_1, the following condition can be obtained:

$$1 + W(j\omega_1)\left[a_{11} - \frac{a_{1n}a_{n1}W(j\omega_n)}{1 + a_{nn}W(j\omega_n)}\right] = 0. \qquad (17.27)$$

Fig. 17.3 Structural
representation of combined
two-frequency parametric
excitation

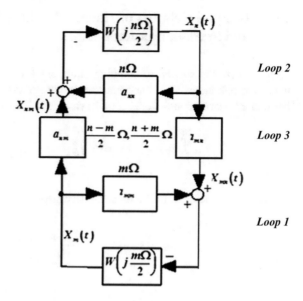

The latter has its structural representation depicted in Fig. 17.3, where $m = 1$, $\omega_1 = \Omega/2$, and $\omega_n = n\,\Omega/2$.

The common structure representation includes two coupled loops of the first parametric resonance. Their excitation leads to the stability loss of the single-frequency solution $x_m(t)$ of Eq. (17.24). Coupling harmonics coefficients (16.6) have the forms

$$a_{nm} = \frac{2j}{T_n} \int_0^{T_n} \sum_{k=-\infty}^{+\infty} a_k e^{jk\Omega(t-\tau)} \sin\frac{m}{2}\Omega t\, e^{-jn\Omega t/2} = a_{\frac{n-m}{2}} e^{-j\frac{n-m}{2}\varphi} - a_{\frac{n+m}{2}} e^{-j\frac{n+m}{2}\varphi}$$

$$a_{mn} = \frac{2j}{T_m} \int_0^{T_m} \sum_{k=-\infty}^{+\infty} a_k e^{jk\Omega(t-\tau)} \sin\frac{n}{2}\Omega t\, e^{-jm\Omega t/2} = a_{-\frac{n-m}{2}} e^{+j\frac{n-m}{2}\varphi} - a_{\frac{n+m}{2}} e^{-j\frac{n+m}{2}\varphi}$$

$$\tag{17.28}$$

$$a_{mm} = -a_m e^{-jm\phi}, \quad a_{nn} = -a_n e^{-jn\phi}, \quad \varphi = \Omega\tau$$

As noted in Chap. 3, they are uncomfortable in use. Indeed, the derivative of the nonlinear function does not exist often, many typical nonlinearities have discontinuities or kinks. Moreover harmonic linearization factors are derived at the stage of oscillation analysis and it is convenient to use them for evaluation of periodic derivative expansion factors. The same way the first parametric resonance circle was evaluated by the first harmonic approximation coefficients with expression (8.19).

17.4 General Relation of Harmonic Linearization Factors and Stationarization Factors

To generalize the system of coupling equations for the harmonic motion components and parameter variation components, we again departure from the following harmonic linearization factors for odd-symmetrical nonlinearity:

$$W_m(A) = \frac{j}{\pi A} \int_0^{2\pi} F(A \sin \psi)e^{-jm\psi}\,d\psi, \quad \psi = \omega t, \quad m = 1, 3, 5, \ldots \quad (17.29)$$

Fourier coefficients for the function derivative are

$$a_0 = \rho_0 \frac{1}{2\pi} \int_0^{2\pi} \frac{dF(A \sin \psi)}{dx}e^{-jk\psi}\,d\psi.$$

$$a_k = \frac{\rho_k}{2j} = \frac{1}{\pi} \int_0^{2\pi} \frac{dF(A \sin \psi)}{dx}e^{-j2k\psi}\,d\psi, \quad k = 1, 2, \ldots \quad (17.30)$$

We first derive the difference

$$
\begin{aligned}
\frac{a_{m-1}}{2} - \frac{a_{m+1}}{2} &= \frac{j}{\pi} \int_0^{2\pi} \frac{dF(A \sin \psi)}{dx}[e^{-j(m-1)\psi} - e^{-j(m+1)\psi}]\,d\psi \\
&= -\frac{2}{\pi} \int_0^{2\pi} \frac{dF(A \sin \psi)}{dx}\sin\psi\, e^{-j2m\psi}\,d\psi \\
&= -\frac{2}{\pi} \int_0^{2\pi} \frac{dF(A \sin \psi)}{dx}\frac{dx}{dA}e^{-j2m\psi}\,d\psi \\
&= 2j\frac{d[AW_m(A)]}{dA}
\end{aligned}
\quad (17.31)
$$

and then the sum

$$\frac{a_{m-1}}{2} - \frac{a_{m+1}}{2} = \frac{j}{\pi} \int_0^{2\pi} \frac{dF(A\sin\psi)}{dx}[e^{-j(m-1)\psi} - e^{-j(m+1)\psi}]d\psi$$

$$= \frac{j}{\pi} \int_0^{2\pi} \frac{dF(A\sin\psi)}{dx}\cos\psi\, e^{-j2m\psi}d\psi$$

$$= \frac{2j}{\pi A} \int_0^{2\pi} \frac{dF(A\sin\psi)}{dx}\frac{dx}{d\psi}e^{-j2m\psi}d\psi \qquad (17.32)$$

$$= \frac{2j}{\pi} \int_0^{2\pi} \frac{d[AW_m(A)]}{dA}e^{-jm\psi}dF(A\sin\psi)$$

$$= \frac{2m}{j}\frac{j}{\pi A} \int_0^{2\pi} F(A\sin\psi)e^{-jm\psi}d\psi = 2jmW_m(A).$$

Now, adding and subtracting (17.31) and (17.32), we arrive at

$$a_{\frac{m-1}{2}} = 2j\left(\frac{m+1}{2}W_m(A) + \frac{A}{2}\frac{dW_m(A)}{dA}\right),$$

$$a_{\frac{m+1}{2}} = 2j\left(\frac{m-1}{2}W_m(A) - \frac{A}{2}\frac{dW_m(A)}{dA}\right).$$

Applying the complex Fourier expansion coefficient notations to the two last relations, the coupling equations are derived as follows:

$$\rho_{\frac{m-1}{2}} = a_{\frac{m-1}{2}}/2j = \frac{m+1}{2}W_m(A) + \frac{A}{2}\frac{dW_m(A)}{dA}, \qquad (17.33)$$

$$\rho_{\frac{m+1}{2}} = a_{\frac{m+1}{2}}/2j = \frac{m-1}{2}W_m(A) - \frac{A}{2}\frac{dW_m(A)}{dA}, \qquad (17.34)$$

here $m = 1,3,5,\ldots$ and the first parametric resonance circle coordinates (8.19) obtained for the single-frequency approximation arise for $m = 1$ assuming the equality $a_0 = \rho_0$. These coupling Eqs. (17.33) and (17.34) enable to get the periodic parameter Fourier expansion coefficients (in other words, the derivative of the non-linear function of a periodic argument) by the known harmonic linearization factors.

To complete this chapter, it should be noted again that the enumeration and consideration of all the harmonic parameter components or linear functions in calculations are unfeasible and unpractical. That is why the assessment of higher harmonic amplitudes seems rational prior to multifrequency analysis. Having compared them to the frequency response modulus of the linear stationary part, one can leave off the smaller order terms. Moreover, by no means all the harmonic components of the expansions affect the evaluation of corrections and a_{nm} harmonic couplings. Thus in the general set of the expansion coefficients of a periodic derivative

$$a_{nm} = \frac{2j}{T} \int_0^T \frac{dF(x)}{dx}\bigg|_{x=x_1} \sin m\omega t \; e^{-jnm\omega t} dt, \qquad (17.35)$$

only two harmonics of numbers $(m-1)/2$ and $(m+1)/2$, $m = 1, 3.5\ldots$, are used in the correction of the harmonic linearization factor of the odd-symmetrical nonlinear characteristic F for $n = 1$ as follows:

$$
\begin{aligned}
a_{1m} &= \frac{2j}{T} \int_0^T \frac{dF(x)}{dx}\bigg|_{x=x_1} \sin m\omega t \; e^{-j\omega t} dt \\
&= \frac{2j}{T} \int_0^T \left[a_{\frac{m+1}{2}} \sin(m+1)\omega t + a_{\frac{m-1}{2}} \sin(m-1)\omega t \right] \sin m\omega t e^{-j\omega t} dt \\
&= \frac{j}{2} \left(a_{\frac{m+1}{2}} + a_{\frac{m-1}{2}} \right)
\end{aligned}
\qquad (17.36)
$$

Chapter 18
Robust Dynamic Systems

18.1 Robustness: Concepts and Definitions

Problem background. There is roughly one and half century history of investigations of sensitivity with respect to different kinds of uncertainties in the description of the system or external disturbances.

In 1876, the analysis by I. Vishnegradskiy opened the chapter of stability analysis in parametrically disturbed systems. By disturbed system, we mean a system in which parameters are uncertain but fixed. The idea was to specify the regions for the characteristic polynomial $G(s)$'s parameters g_0, g_1, ..., g_{n-1} where the system is asymptotically stable, or has a given distribution of characteristic equation roots. Or, what it same, to define these regions in the state-space system parameters a_{ij} that are the components of the state-space matrix A. Then, so-called D-discretization method suggested by U. Neymark in 1948 became popular. The method designed the stability region on parametric plane (the space of two adjustable parameters). Working with state-space representations, the stability analysis under **parametric** disturbances can be performed by Lyapunov method. In 1932, Nyquist suggested a frequency stability criterion in order to investigate the stability of parametrically disturbed linear stationary systems, or systems, whose structure is unknown but fixed. Another criterion suggested by V. Popov in 1958 provided the sufficient conditions of absolute stability by the specification on the linear part of the system.

But no stability criteria can tell how close is the stable system to instability and how it is sensitive to the variations (disturbances) of its parameters and structure. Just this line was set up in 1937 by A. Andronov and L. Pontryagin who introduced the fundamental concept of *a robust system*. So, the system was named *robust* if the topological structure of a phase plane did not change under small variations of the describing differential equations. Later on, E. Rozenvasser added to this work the sensitivity analysis of control systems. At present, the robustness is in fact one of the mathematical model quality standards of nonsingular physical systems and processes. In the beginning of 1980s, Zames described the class of *robust* control systems characterized by the ability to keep their stability under sufficiently large

© Springer International Publishing AG, part of Springer Nature 2017
L. Chechurin and S. Chechurin, *Physical Fundamentals of Oscillations*,
https://doi.org/10.1007/978-3-319-75154-2_18

but bounded description disturbances. Exactly the facts that the disturbances were presumed to be nonlocal and the stability but not the phase portrait was chosen as a key indicator marked the opening of new subchapter in robustness studies. By that time, several researchers the results on parametric robustness: in 1978, V. Kharitonov published his famous theorem on interval characteristic polynomial stability.

The principal milestone of robust feedback system synthesis studies was the reduction of the robust control problem under structural disturbances to the standard feedback frequency synthesis with the Hardy space or H_∞ norm quality index. Thanking to the progress in mathematical treatment of the problem, the appropriate indexes and optimization approaches became available in 90s. Moreover, the theoretical approach enabled solving more complicated problems within standard optimal feedback synthesis like various multi-criterial control synthesis, for example, for robustness and performance, etc. The results are widely presented in the works by K. Glover, J. Doyle, H. Kwakernaak, B. Francis, A. S. Pozdniak, A. A. Pervozvansky, E. A. Barabanov, N. A. Barabanov, V. A. Yakubovich, and many others.

Problem definition. In the practice of control design, we often deal with situation when the parameters and even structure of the object are uncertain. However, the bounds for uncertainty are available. There are always measurement errors of, for example, mass, stiffness, inductance, etc. Moreover, the operating conditions can assume certain range of parameter values: for example, a truckload could belong to the range from zero to its maximum capacity, turbine rotor could rotate at various operation rates, etc. The structure can be known to certain extent. For example, the rotor can be modeled as a rigid body while some unmodeled dynamics like high frequency oscillation modes can be observed as dynamic object uncertainties. In the same manner, the object itself can have scheduled structure change with respect to certain situation. For example, a cargo can be rigid or liquid. The latter case means that the fluid dynamic model should be applied.

In this case, we have to make sure that the designed controller ensures closed loop performance for any of the parameter values or structure from the given set. They typically consider two cases.

Parametric perturbation robustness. Parametric perturbations are obviously a particular case of structural ones. But there are special results for the former and they yield simple and less conservative assessments. The tools based on Kharitonov's theorem seem to be the most appropriate. The main idea of the theorem is the following. Let the characteristic polynomial of the closed-loop system is known purpose. Let Q be a region of uncertainty in the description of characteristic polynomial coefficients

$$\{Q \equiv g: \quad \underline{g}_i \leq g_i \leq \overline{g}_i\}.$$

Thanking to Kharitonov theorem, we do not need to check stability for any possible polynomial from the region Q region. The system is stable if and only if the following four polynomials are stable:

$$\alpha_1(s) \equiv \underline{g}_0 + g_1 s + \overline{g}_2 s^2 + \overline{g}_3 s^3 + \underline{g}_4 s^4 + \underline{g}_5 s^5 + \cdots$$
$$\alpha_2(s) \equiv \underline{g}_0 + \overline{g}_1 s + \overline{g}_2 s^2 + g_3 s^3 + \underline{g}_4 s^4 + \overline{g}_5 s^5 + \cdots$$
$$\alpha_3(s) \equiv \overline{g}_0 + \overline{g}_1 s + \underline{g}_2 s^2 + \underline{g}_3 s^3 + \overline{g}_4 s^4 + \overline{g}_5 s^5 + \cdots$$
$$\alpha_4(s) \equiv \overline{g}_0 + \underline{g}_1 s + \underline{g}_2 s^2 + \overline{g}_3 s^3 + \overline{g}_4 s^4 + \underline{g}_5 s^5 + \cdots.$$

There are many modification of the theorem. Thus, in [16], we developed the sufficient approximate stability conditions for time-varying polynomial coefficients $g_i(t)$ based on Kharitonov theorem and stationarization method.

Structure perturbation robustness implies the system's ability to maintain the required functioning under perturbations related to changing the order to the object's transfer function. However, any variation including structural can be projected on the frequency response domain. That is why frequency approach to robustness analysis seems to be the most appropriate tool. Small gain theorem is the best illustration for this approach. Let the open system transfer function $W(s)$ be stable. Nyquist criterion says that the necessary and sufficient condition for the closed system to be stable is that the hodograph $W(j\omega)$ does not encircle the point $(-1, j0)$ on the complex plane. The theorem states the following obvious corollary: it is sufficient for the stability to have the hodograph inside the unit circle, i.e.,

$$|W(j\omega)| < 1, \forall \omega, \quad \text{or}$$
$$\max_{\omega} |W(j\omega)| < 1$$

In other words, the open-loop system passivity is sufficient to provide the stability of the closed-loop system.

An important generalization can take place with respect to multi-input multi-output (MIMO) systems where $W(j\omega)$ is the matrix of rational transfer functions. But what could play the role of Nyquist hodograph for MIMO transfer matrix? However, the scalar small gain theorem can be generalized by the H_∞ norm or the Hardy space norm: the closed system is stable if

$$||W(j\omega)||_\infty < 1.$$

It is assumed here that

$$||G(j\omega)||_\infty = \text{ess} \sup_{\omega} \lambda(W(j\omega)),$$

where $\lambda(a)$ is a maximum singular value of the matrix a.

This approach can be used to evaluate the stability robustness of stationary control systems when the object description has uncertainties.

Indeed, let us assume that the control system presented in Fig. 18.1 is considered and the object transfer function undergoes perturbations. This can be written down, e.g., in the form (see Fig. 18.2a).

Fig. 18.1 Control system
with perturbed object

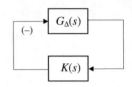

Fig. 18.2 Additive plant
perturbation control scheme
(a) and its equivalent scheme
(b)

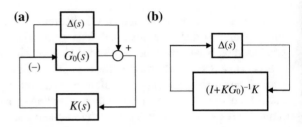

$$G_\Delta(s) = G_0(s) + \Delta(s), \tag{18.1}$$

where $G_0(s)$ is a nominal object, $\Delta(s)$ are structure perturbations. Let also the uncertainty $\Delta(s)$ belong to the Q_ε class which means that

$$\|\Delta(s)\|_\infty \leq \varepsilon, \quad \varepsilon > 0$$

and the functions $G(s)$ and $G_0(s)$ have the same number of unstable poles. Obviously, a structural transformation can reduce the problem analysis to the scheme where small gain theorem works (see Fig. 18.2b).

Thus, it may be stated that the sufficient condition for the stability of the closed system for any $\Delta(s) \in Q_\varepsilon$, is $K(s)$ stabilizes the nominal object and the inequality

$$\left\|(I + K(s)G_0(s))^{-1} K(s)\right\|_\infty \leq \frac{1}{\varepsilon} \tag{18.2}$$

holds.

Robustness criterion has extensions for the cases when the uncertainty is modeled as multiplicative or fractional (in numerator or/and denominator) perturbation of the object's transfer function. Moreover, if the frequency-dependent upper boundary function of the transfer function perturbations is known, or in other words, a scalar transfer function $V(\omega)$ is given so that the condition

$$\lambda(\Delta j\omega) \leq \varepsilon \, |V(j\omega)| \tag{18.3}$$

holds for any frequency ω, the robustness criterion takes a bit more conservative formulation

$$\left\|(I + K(s)G_0(s))^{-1} K(s)V(s)\right\|_\infty \leq \frac{1}{\varepsilon}. \tag{18.4}$$

Fig. 18.3 Standard problem
of robust filtration

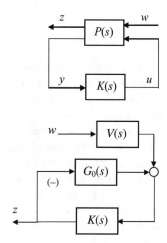

Fig. 18.4 Auxiliary problem
for robust stability synthesis

If the system satisfies inequality (18.4), it is stable under any of the above perturbations. And if the inequality fails at some of the frequencies, the stability cannot be guaranteed.

Robust synthesis. If the left part of inequality (18.2) or (18.4) is regarded as a synthesis criterion for the feedback regulator $K(s)$, one can observe this problem as optimization. Indeed, we wonder if we can design a controller that minimizes this criterion, or in other words, the one to maximize the stability margin. The robust control theory provides instruments to address this problem. The optimal design procedures with robust stability criterion are rather complicated and far beyond the focus of this book. Let us consider just the basic idea. The layout for the feedback regulator $K(s)$ design depicted in Fig. 18.3 is the departure point. The measured output y is used by the regulator to minimize the power of the controlled output signal z, subject the external disturbance w has power bounded but the worst possible spectrum.

That problem is called a standard H_∞ synthesis problem or an optimal robust filtration problem. If the description of the system is given in state-space form, the design of H_∞-norm optimal (in general case, suboptimal) controller is available. It is based on the iterative solution of two Lurie–Ricatti equations and made its way to the commercial software products, for example, MATLAB (MATLAB/Robust toolbox).

But the most important fact is that the internal perturbation robustness problem can be reduced to the standard H_∞ synthesis problem. Indeed, let us consider the auxiliary scheme depicted in Fig. 18.4. The transfer function from w to z is exactly the left part of (18.4). Thus, having designed a controller for optimal robust filtration, we can use it for robust stability.

It remained to be noted that the auxiliary problem is a standard robust filtration problem as illustrated in Fig. 18.4 where the object transfer function is

$$P(s) = \left[\begin{array}{c|c} 0 & V(s) \\ \hline I & G_0(s) \end{array} \right].$$

It is worth emphasizing again that though the regulator to be constructed as a result of solving a standard problem is expected to be the best, it is essential to test if it satisfies the condition (18.4). If it does not hold, we are not able to guarantee robust stability under given uncertainty level (18.3).

Thousands and thousands of publications on robust synthesis develop the theory, applications, and algorithms. They reduce the conservatism, combine various criteria, and approach time-variant and nonlinear systems. At the same time, the robust design approach faces serious difficulties in practice since the results are very mathematically involved and conservative. In this respect, approximate solutions and assessments would be of big help for the engineers. The following chapters use the method of stability analysis based on first frequency approximation. We emphasize that one method covers a wide spectrum of systems: time-variant, nonlinear, discrete, with different types of linear part, etc. We also are able to approach new problems such as the analysis of robust stability with respect to dynamic parameter or structure variation or with respect to dynamic coordinate disturbances of perturbations.

18.2 Robustness with Respect to Periodic Parameter Variation Bounded in Average

The periodic nonstationary system in which an unbounded in magnitude positive periodic parameter has a period-averaged constant value of $a/2$ is considered. Let the parameter have only two extreme values: zero and a_{max}, as shown in Fig. 18.5,

$$a_{max} = \frac{a}{2\gamma}. \tag{18.5}$$

Let us define the conditions of parametric resonance for parameter variations for the specified class. We need to design the family of parametric resonance circles. First of all, the center of the circles (that is the average value) is the same for the whole

Fig. 18.5 Periodic parameter change, bounded in average

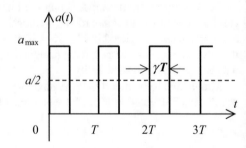

Fig. 18.6 First parametric
resonance conditions

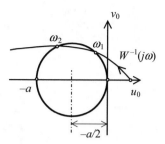

class and it is at the distance of $a/2$ from the origin. The first harmonic amplitude is
evaluated as

$$
a_1 = \frac{2j}{T} \int_0^T a(t)e^{-j\Omega t}\,dt = \frac{ja}{\gamma T} \int_0^{\gamma T} e^{-j\Omega t}\,dt = \frac{a}{2\gamma\pi}\left(1 - e^{-j2\pi\gamma}\right). \tag{18.6}
$$

From which, we can define the radius of the first parametric resonance circles as

$$
r = \left|\frac{c_1}{2j}\right| = \frac{a}{4\pi\gamma}\left|1 - e^{-j2\pi\gamma}\right| = \frac{a}{2\sqrt{2}\pi\gamma}\sqrt{1 - \cos 2\pi\gamma} = \frac{a}{2}\frac{\sin\pi\gamma}{\pi\gamma}. \tag{18.7}
$$

The circle radius reaches its maximum value in the range $0 \le \gamma \le 1$ when $\gamma = 0$,
i.e.,

$$
r_{\max} = a/2. \tag{18.8}
$$

Since the concentric circles take place, their envelope is the circle of maximum
radius (18.8):

$$
W(j\phi) = \frac{a}{2}\left(1 - e^{-j\phi}\right) \tag{18.9}
$$

The circle crosses the origin, as illustrated in Fig. 18.6 in the inverse Nyquist
hodograph plane; the system becomes unstable if Nyquist hodograph enters the
circle.

The circle $-W^{-1}(j\varphi)$ degenerates into the vertical straight line crossing the point
$(-a^{-1}, j0)$ in the regular Nyquist hodograph plane. The stability condition over the
class of the parameter variations restricted in the average, or robustness condition,
looks like (see Fig. 18.7):

$$
\mathrm{Re}\,W(j\omega) \ge -a^{-1}. \tag{18.10}
$$

The analysis shows that the most dangerous parameter variation form with respect
to system stability is δ-function.

Fig. 18.7 Parametric
resonance conditions in
Nyquist hodograph plane

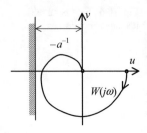

18.3   Robustness with Respect to Periodic Parameter Magnitude

Let the periodic parameter $a(t)$ have an arbitrary variation bounded in magnitude as

$$|a(t)| \leq a/2. \tag{18.11}$$

There is sufficient Bongiorno stability criterion to a nonstationary system with restriction (18.11) that imposes the following restriction to the stationary part of the system:

$$\left|W^{-1}(j\omega)\right| \geq a/2 \tag{18.12}$$

The geometrical interpretation can be easily given, see Fig. 18.8.

Moreover, the parameter varies within the limits of $-a/2 \leq a(t) \leq +a/2$. But if the parameter varies in the range of $0 \leq a(t) \leq a$, the circle and the inverse frequency response of Fig. 18.8 have to be shifted by $a/2$ to the left along the real axis. In that case, the circle is the same as the circle (18.9) and the stability condition is equal to the condition (18.10), i.e.,

$$\operatorname{Re}W(j\omega) \geq -a^{-1}. \tag{18.13}$$

It should be stressed that the coincidence is only formal. In spite of the fact that the mean parameter values are the same in both cases, their periodic components are

Fig. 18.8 Bongiorno
stability criterion

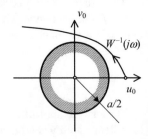

Fig. 18.9 Amplitude-
limited parameter
cycling

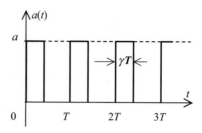

different. Furthermore, condition (18.10) is approximate, while condition (18.13) is
strict and sufficient.

Let us assume that the parameter jumps in the limits of $0 \le a(t) \le a$, as represented
in Fig. 18.9, and calculate the first harmonic and the first parametric resonance
circle radius again to compare Bongiorno stability criterion and the first harmonic
approximation. Then, the constant parameter component and the first harmonic are

$$a_0 = \frac{c_0}{2j} = a\gamma, \quad 0 \le \gamma \le 1 \tag{18.14}$$

$$c_1 = \frac{2ja}{T} \int_0^{\gamma T} e^{-j\Omega t} dt = \frac{a}{\pi}(1 - e^{-j2\pi\gamma}). \tag{18.15}$$

Expressions (18.14) and (18.15) define the parametric resonance circle radius

$$r = \left| \frac{c_1}{2j} \right| = \frac{a}{2\pi} \left| 1 - e^{-j2\pi\gamma} \right| = \frac{a}{\pi} \sin \pi\gamma. \tag{18.16}$$

With a fixed pulse ratio γ, the parametric resonance circle equation and the stability loss condition of a periodic system in the inverse Nyquist hodograph plane
$[u_0 = \mathrm{Re}W^{-1}(j\omega), v_0 = \mathrm{Im}W^{-1}(j\omega)]$ become

$$(u_0 + \gamma)^2 + v_0^2 = \frac{\sin^2 \pi\gamma}{\pi^2} \tag{18.17}$$

We need to define the envelope for the family of circles (18.17) to find the system
stability loss conditions over the class of a modulus-bounded parameter variation,
in other words for any γ from the interval $[0, 1]$. The derivative of Eq. (18.17) with
respect to γ is

$$u_0 = -\gamma + \frac{\sin 2\pi\gamma}{2\pi}. \tag{18.18}$$

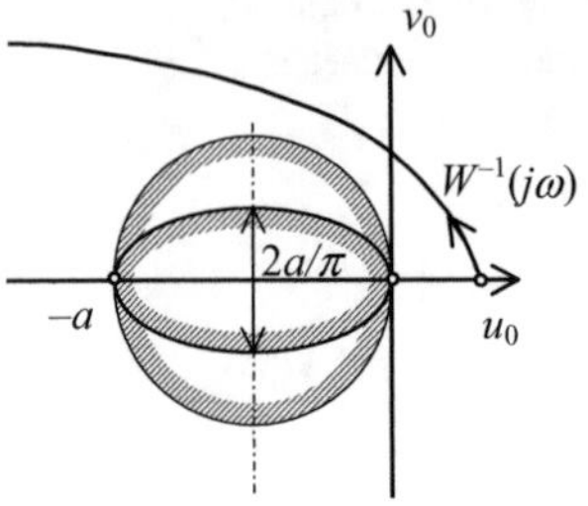

Fig. 18.10 Frequency
response envelope for
modulus-bounded parameter
variation

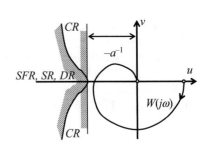

Fig. 18.11 Robustness
regions in Nyquist
hodograph plane

Then condition (18.17) gives the solution

$$v_0 = \pm \frac{\sin^2 \pi \gamma}{\pi} \qquad (18.19)$$

Expressions (18.18) and (18.19) describe the oval. It envelops the circles (18.17) in the inverse Nyquist hodograph plane, shown in Fig. 18.10.

If the inverse frequency response of the stationary part of the system does not enter the envelope, the periodically nonstationary system is robust over the class of periodic parameter variations. Circle (18.9) is plotted in Fig. 18.10 for comparison. The system is robust in the first harmonic approximation over the class of parameter variation bounded in average if the stationary part frequency response does not enter the circle. And, as it was shown, it is stable according to the sufficient Bongiorno criterion. Figure 18.11 illustrates the same conditions on the plane of Nyquist hodograph.

It is interesting to note that the Nyquist hodograph plane is divided into three regions. The regions of single-frequency, sum, and difference resonances denoted as SFR, SR, and DR, correspondingly. They are located to the left from the envelope; the stable region is to the left from the vertical straight line and the region of combined multifrequency oscillations (CR) is between the envelope and the vertical straight line.

18.4 Time-Variant System Robustness with Respect to Parameter Variation Frequency and Static Gain

Let us consider the periodically nonstationary system

$$G(s)x(t) + kH(s)a(t)x(t) = 0,   \tag{18.20}$$

where the static gain k is explicitly shown so that $G(0) = H(0) = 1$ and the positive single-frequency harmonically variable parameter

$$a(t) = a_0 + a \sin \Omega t,   0 \le a < a_0.   \tag{18.21}$$

We need to define the intervals of frequencies Ω and gain coefficients variation in which parametric resonance excitation is possible. The first parametric resonance circle is given in Fig. 18.12 on the plane of the inverse frequency characteristic $W^{-1}(j\omega) = H(j\omega)/G(j\omega)$.

As it follows from Fig. 18.12, the family of circles $ka(t)$ belongs to the sector of angle

$$\psi_P = \pm arctg \frac{r}{\sqrt{a_0^2 - r^2}}.   \tag{18.22}$$

Here, subscript P implies that this is the phase shift of parameter. Thus, if the inverse frequency characteristic does not enter the sector (18.22) of the left semiplane, the dynamic system is robust over any positive gain value.

Since the angle ψ_P and phase of the frequency characteristic ψ_L make a flat angle, the following ratio can be written down:

$$-tg\psi_L = -\frac{\text{Im} W^{-1}(j\omega)}{\text{Re} W^{-1}(j\omega)} = \frac{r}{\sqrt{a_0^2 - r^2}}.   \tag{18.23}$$

Here, subscript L implies that this is the phase shift of the stationary open-loop part. The solution $\omega_{\min}$ derived from ratio (18.23) is in general the minimal fre-

Fig. 18.12 Parametric resonance conditions in inverse Nyquist hodograph plane

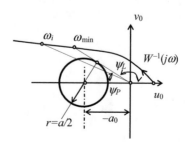

quency of parametric oscillations at the parameter variation frequency $\Omega_{min} = 2\omega_{min}$. Accordingly, if the inverse frequency characteristic falls inside sector (18.22) at the frequency ω_{min} and does not leave it, as depicted in Fig. 18.12 for an example, then the periodically nonstationary system (18.20) is robust over the frequency range $0 \leq \Omega \leq \omega_{min}$.

Finally, we can define the gain boundaries for the parametric resonance excitation for any point $\omega_I > \omega_{min}$ belonging to the sector (18.22) of the inverse frequency response. First of all, the ratio has to be determined

$$\lambda = -\frac{\mathrm{Im}\,W^{-1}(j\omega_i)}{\mathrm{Re}\,W^{-1}(j\omega_i)}. \tag{18.24}$$

Then intersections (18.25) and (18.26) of the circle and the ray $0\omega_i$ or $(u_{0i}^{\pm}, v_{0i}^{\pm})$, see Fig. 18.12, intersection modulus (18.27), and ω_i point modulus are defined as follows:

$$u_{0i}^{\pm} = \frac{-a_0 \pm \sqrt{r^2 - \lambda^2(a_0^2 - r^2)}}{1 + \lambda^2}. \tag{18.25}$$

$$v_{0i}^{\pm} = -\lambda u_{0i}^{\pm}. \tag{18.26}$$

$$M_{0i}^{\pm} = \sqrt{1 + \lambda^2}\,\left|u_{0i}^{\pm}\right|. \tag{18.27}$$

$$\left|W^{-1}(j\omega_i)\right| = \sqrt{\mathrm{Re}^2\,W^{-1}(j\omega_i) + \mathrm{Im}^2\,W(j\omega_i)}.$$

The excitation interval can now be obtained from

$$\frac{\left|W^{-1}(j\omega_i)\right|}{M_{0i}^{-}} \leq k \leq \frac{\left|W^{-1}(j\omega_i)\right|}{M_{0i}^{+}}. \tag{18.28}$$

The robustness area of the periodically nonstationary system at the parameter variation frequency of $\Omega_i = 2\omega_i$ belongs to outside of the interval (18.28) if Hurwitz conditions are satisfied for the stationary part of the system with the mean parameter value a_0.

It is clear that the graphical analysis like that of Fig. 18.12 for the above-considered problems is not difficult. The same robustness conditions can be derived on the Nyquist hodograph plane.

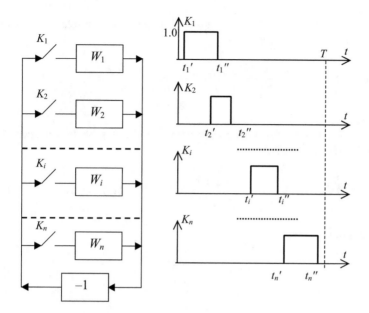

Fig. 18.13 Variable structure system

18.5 Robustness of Periodically Nonstationary Systems with Respect to Structure Variation

When the structure of dynamic systems changes once or changes aperiodically or varies slowly, the nonstationary system is considered to be structure robust if its initial, all intermediate, and final structures are Hurwitzian. However, with a periodic variation of the structure, the intermediate Hurwitzian stability of each realization is not sufficient property to predict system robustness.

Let us consider the system depicted in Fig. 18.13. Its structure changes n times per period in a step manner by K keys locked in turn, which is illustrated in the timing diagrams next to the block scheme. A stepwise structure change is not so rare situation in engineering practice. An electrical motor, for example, can be considered as an element of the periodic structure if the rotor winding is periodically switched off. There are examples of similar structural changes in mechanics, economics, etc.

The variable structure system shown in Fig. 18.13 is a periodically nonstationary synchronous multiparameter system. According to the diagrams, the system parameters, i.e., each of the n keys takes a unit value in the time range from t_i' to t_i'' as

$$
\begin{aligned}
a_1(t) &= 1, \quad 0 = t_1' < t < t_1'' \\
a_2(t) &= 1, \quad t_2' < t < t_2'' \\
a_n(t) &= 1, \quad t_n' < t < t_n''
\end{aligned}
\tag{18.29}
$$

Each of the parameters has the transfer function based on relations (18.14) and (18.15) at $a = 1$ as follows:

$$W_i(j\varphi) = \gamma_i + \frac{(1 - e^{-2\pi\gamma_i})}{\pi}e^{-j(\varphi+\lambda_i)} = \gamma_i - \frac{\sin\pi\gamma_i}{\pi}e^{-j(\pi\gamma_i+\lambda_i+\varphi)}, \qquad (18.30)$$

where $\lambda_i = \Omega t_i'$, $\gamma_i = \frac{t_i''-t_i'}{T}$.

The first parametric resonance excitation condition is obtained according to Fig. 18.13 with $s = j\Omega/2$ as

$$\sum_{i=1}^{n}\gamma_i W_i(s) - e^{-j\varphi}\sum_{i=1}^{n}\frac{\sin\pi\gamma_i}{\pi}e^{-j(\pi\gamma_i+\lambda_i)}W_i(s) = -1. \qquad (18.31)$$

In the notations

$$W_{0e}(j\omega) = \sum_{i=1}^{n}\gamma_i W_i(j\omega) \qquad (18.32)$$

$$W_{1e}(j\omega) = \sum_{i=1}^{n}\frac{\sin\pi\gamma_i}{\pi}e^{-j(\pi\gamma_i+\lambda_i)}W_i(j\omega) \qquad (18.33)$$

condition (18.31) takes the form

$$W_{1e}^{-1}(j\omega)[1 + W_{0e}(j\omega)] = e^{-j\varphi} \qquad (18.34)$$

or eliminating φ,

$$|W_{1e}(j\omega)| = |1 + W_{0e}(j\omega)|. \qquad (18.35)$$

It is not difficult to obtain frequency responses (18.32) and (18.33) by manipulating with initial frequency characteristics $W_i(j\omega)$ on the frequency response plane: scaling, rotating, and summating.

In practice, K keys are most often commuted inside the period and not at arbitrary time points, as specified in relations (18.29). Instead, the keys are locked for equal time period τ_{cl}, at the beginning of every uniformly spread time interval τ_0. In other words, we can formalize it as

$$\tau_{cl} = t_i'' - t_i' = const, \quad \gamma_i = \gamma = \tau_{cl}/T, \quad \tau_0 = T/n, \quad \tau_{cl} \leq \tau_0, \quad \lambda_i = 2\pi(i-1)/\blacksquare \qquad (18.36)$$

and equivalent frequency responses (18.32) and (18.33) become

(a)   **(b)**

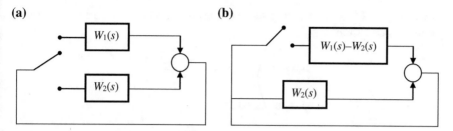

Fig. 18.14 Example. Variable structure system (**a**) and its equivalent (**b**)

$$W_{0e}(j\omega) = \gamma \sum_{i=1}^{n} W_i(j\omega), \quad W_{1e}(j\omega) = \frac{\sin \pi \gamma}{\pi} e^{-j\pi\gamma} \sum_{i=1}^{n} W_i(j\omega) e^{-j2\pi(i-1)/n}$$

$$(18.37)$$

In this case, the constant factor outside the summation symbol in condition (18.31) is convenient to attribute to the circle $\exp(-j\varphi)$ and rewrite conditions (18.34) and (18.35) in the reduced forms

$$1 + W_{0e}(j\omega) = \frac{\sin \pi \gamma}{\pi} e^{-j(\varphi+\pi\gamma)} W'_{1e}(j\omega), \tag{18.38}$$

$$|1 + W_{0e}(j\omega)| = \frac{\sin \pi \gamma}{\pi} |W'_{1e}(j\omega)|, \tag{18.39}$$

where

$$W'_{1e}(j\omega) = \sum_{i=1}^{n} W_i(j\omega) e^{-j2\pi(i-1)/n}. \tag{18.40}$$

Finally, for the simplest but widespread case of the two switched structures ($n = 2, \gamma = 0.5$)

$$W_{0e}(j\omega) = 0.5[W_1(j\omega) + W_2(j\omega)], \quad W'_{1e}(j\omega) = [W_1(j\omega) - W_2(j\omega)]. \tag{18.41}$$

Example. We consider a feedback system with variable two-element structure presented in Fig. 18.14a. There are two approaches to analyze the stability of the system. The first approach is to reduce the system to the form shown in Fig. 18.14b by simple structural transformation, then to replace the key by its equivalent transfer function in the first harmonic approximation in the form of (18.30), and then apply Nyquist criterion. The second approach is the direct application of expressions (18.41).

In both cases, the balance condition is obtained as

$$(W_1(s) - W_2(\underline{s}))W_i(j\varphi) + W_2(s) = -1,$$

where $W_1(s)$, $W_2(s)$ are the transfer functions of the stationary part of the system under two conditions and

$$W_i(j\varphi) = \gamma - \frac{e^{-j\pi\gamma - j\varphi} \sin \pi\gamma}{\pi}.$$

The last equality can be represented as

$$-W_i(j\varphi) = \frac{1 + W_2(s)}{W_1(s) - W_2(s)}.$$

The condition can be given a clear geometric interpretation.

Let the variable structure system shown in Fig. 18.14a be considered. To observe the excitation of the first parametric resonance at frequency ω that is half of structure change frequency $\Omega/2$, it is necessary that the hodograph of the transfer function

$$W(s) = \frac{1 + W_2(s)}{W_1(s) - W_2(s)}$$

enters the circle $-W_i(j\varphi)$ of the first parametric resonance.

To provide a numerical example, we assume that

$$W_1(s) = \frac{s + 0.2}{s^2 + 0.6s + 0.05},$$

$$W_2(s) = \frac{-s - 0.2}{s^2 + 0.6s + 0.05}.$$

A program according the balance equations calculated the first parametric resonance instability region boundaries for various γ. Then, simulation in MATLAB 6 Simulink tested the obtained results, see Fig. 18.15.

Figure 18.16 displays the obtained excitation region in the coordinates $1/T$, γ (the key lock frequency is the ratio of the $W_1(s)$ lock time to the total lock time).

The excitation region of the system in Fig. 18.14a is neither the integration nor intersection of the excitation regions of separate pulse systems with $W_1(s)$ and $W_2(s)$.

In particular, the frequency instability region derived by simulation for $\gamma = 0.5$ is [0...0.12] Hz and the same obtained by analytical instability criterion above is [0...0.15] Hz.

18.6 Coordinate Robust Nonlinear Systems

Unlike linear systems, nonlinear ones can have stable processes for some ranges of input signals and lose their stability for others. In this respect, the problem formulation of processes stability in nonlinear dynamic systems is legitimate in some class

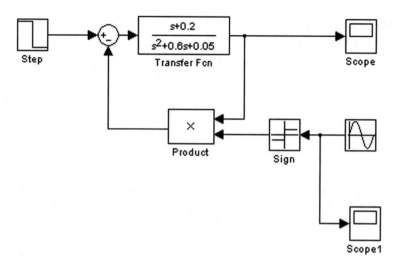

Fig. 18.15 Example. Simulink experiment scheme

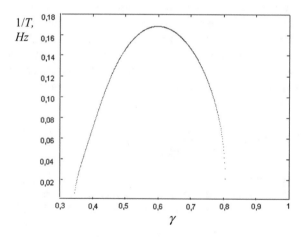

Fig. 18.16 Instability region boundary for variable structure system

of nonlinear characteristics or in a given range of the amplitudes and frequencies of the oscillations or external actions. That is to say, the problem has to be defined with respect to the robustness of nonlinear systems in some class of nonlinear character-istics or in some range of coordinates.

Naumov–Tsipkin sufficient criterion of the sector stability of nonlinear system processes states that a nonlinear system has stable processes if the condition

$$\mathrm{Re}\,W(j\omega) \geq -1/k \tag{18.42}$$

Fig. 18.17 Nonlinearity and its derivative restricted inside the sector

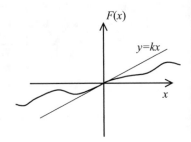

holds in the frequency range of $0 \leq \omega < \infty$ and the single-valued nonlinearity $F(x)$ and its derivative belong to the sector $[0, k]$ as illustrated in Fig. 18.17.

The variable parameter is restricted by the criterion conditions:

$$0 \leq a(t) = \frac{dF[x(t)]}{dt} \leq k. \qquad (18.43)$$

As we can see, the conditions of Bongiorno criterion (18.13) and Naumov–Tsipkin criterion (18.42) are the same at $k = a$, so do their geometric illustrations in the planes of both the direct and inverse frequency characteristics. Interestingly, both criteria admit the formulation given by A. Reshetilov: if the phase difference between the frequency characteristics of open and closed linear part of the system does not exceed $\pi /2$ for all the frequencies, the nonlinear system processes are stable in the class $(0, k)$. Indeed, the condition

$$\left| \arg W(j\omega) - \arg \frac{W(j\omega)}{1 + kW(j\omega)} \right| = \left| \arg k^{-1} + W(j\omega) \right| \leq \frac{\pi}{2} \qquad (18.44)$$

agrees with condition (18.43).

Let us proceed to the comparison of the sufficient and first harmonic approximation process stability conditions and consider again the saturation-like nonlinearity

$$F(x) = \begin{cases} kx, & |x| < b \\ kb, & |x| > b \end{cases} \qquad (18.45)$$

Its derivative can have only two values of $(0, k)$ at the sector boundaries. A harmonic linearization factor is

$$W(A) = k \left(\gamma + \frac{\sin \pi \gamma}{\pi} \right), \quad \gamma = \frac{2}{\pi} \arcsin \frac{b}{A} \qquad (18.46)$$

Then, the center and radius of the first parametric resonance circles are

$$a_0 = k\gamma,$$

$$r = k\frac{\sin \pi \gamma}{\pi}, \tag{18.47}$$

correspondingly. Unlike expressions (18.14) and (18.16), the off-duty factor γ in expressions (18.47) depends on the oscillation amplitude A at the nonlinear element input. The factors coincide in all other respects at $k = a$ and so do their envelopes (18.18) and (18.19) and also the illustrations in Figs. 18.20 and 18.21.

The first harmonic approximation method can be applied to any nonlinearity of general type and define the envelope for process stability loss regions or approximate process stability regions in a given range of oscillation amplitudes and frequencies. The envelopes in their general form can be expressed in terms of harmonic linearization factors. A general equation of negative parametric resonance circles can be written down in the form

$$(u_0 + a_0)^2 + v_0^2 = r^2, \tag{18.48}$$

where the center and the radius depend on harmonic linearization factors (8.19) as

$$a_0 = W(A) + 0.5A\dot{W}(A)$$
$$r = 0.5A\left|\dot{W}(A)\right|. \tag{18.49}$$

Then

$$u_0 = -a_0 + r\frac{\dot{r}}{\dot{a}_0} \tag{18.50}$$

as a result of differentiation of Eq. (18.48) with respect to A.

The second envelope equation is also found from the same equation as follows:

$$v_0 = \pm\sqrt{r^2 - (u_0 + a_0)^2} = \pm\frac{r}{\dot{a}_0}\sqrt{\dot{a}_0^2 - \dot{r}^2}. \tag{18.51}$$

The derivatives of the center and the radius arise from dependences (18.49) as

$$\dot{a}_0 = 0.5[3\dot{W}(A) + A\ddot{W}(A)]$$
$$\dot{r} = 0.5[\dot{W}(A) + A\ddot{W}(A)]. \tag{18.52}$$

Thus taking into account expressions (18.52), the envelope described by Eqs. (18.50) and (18.51) makes the oscillation stability loss zone of the nonlinear system on the inverse Nyquist hodograph frequency region for any specific form of nonlinearity. Several simple examples are given below.

Example 1. The following results have been obtained for the nonlinear element $F(x) = x^2 \, \text{sgn} \, x$:

Fig. 18.18 Example 1.
Robustness sector

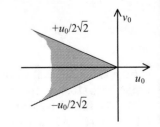

Fig. 18.19 Example 2.
Nonlinear characteristic

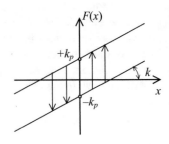

$$W\,(A) = 8A/3\pi,$$
$$a_0 = 4A/\pi, \quad r = 4A/3\pi.$$

Then

$$a_0 = 4/\pi, \quad r = 4/3\pi$$

and Eqs. (18.50) and (18.51) yield

$$u_0 = -32/9\pi,$$
$$v_0 = \pm 8\sqrt{2}A/9\pi.$$

The envelope

$$v_0 = \pm u_0/2\sqrt{2}$$

is the sector between two straight lines in the left half-plane depicted in Fig. 18.18.
By the way, an envelope like that, or $v_0 = \pm u_0/\sqrt{3}$, is obtained for the characteristic
$F(x) = x^3$.

Example 2. Let us consider the nonlinearity shown in Fig. 18.19 which is used to
adjust dynamic system processes because it provides the leading loop and positive
phase shift. The coefficients can be obtained as follows:

$$W\,(A) = k + j4k_p/\pi A^2, \quad a_0 = k, \quad r = 4k_p/\pi A^2.$$

Fig. 18.20 Example 2.
Harmonic linearization

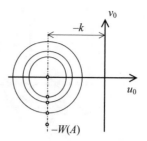

The circles have the common center located at the point $(-k, j0)$ and their radiuses are equal to the imaginary part of harmonic linearization factor (16.28). The circle diameters, as A decreases, become larger, so the circle corresponding to the minimal admissible amplitude of stable oscillations serves as an envelope (Fig. 18.20).

In cases when the circle center is off the real axis, the circle equation

$$(u_0 + a_0')^2 + (v_0 + a_0'') = r^2, \quad a_0 = a_0' + ja_0'' \tag{18.53}$$

has to be used instead of (18.48). Repeating the above steps, the following envelope equations can be obtained:

$$u_0 - a_0' = \frac{r}{\dot{a}_0'^2 + \dot{a}_0''^2}\left(\dot{r}\dot{a}_0' \pm \dot{a}_0''\sqrt{\dot{a}_0'^2 + \dot{a}_0''^2 - \dot{r}^2}\right),$$

$$v_0 - a_0'' = \frac{r}{\dot{a}_0'^2 + \dot{a}_0''^2}\left(\dot{r}\dot{a}_0'' \pm \dot{a}_0'\sqrt{\dot{a}_0'^2 + \dot{a}_0''^2 - \dot{r}^2}\right), \tag{18.54}$$

where the used notations are expressed through the complex harmonic linearization factors as follows:

$$r = 0.5A\left|\dot{W}(A)\right|, \quad a_0' = \mathrm{Re}\left[W(A) + 0.5A\dot{W}(A)\right], \quad a_0'' = \mathrm{Im}\left[W(A) + 0.5A\dot{W}(A)\right]$$
$$\dot{a}_0' = \mathrm{Re}\left[1.5\dot{W}(A) + 0.5A\ddot{W}(A)\right], \quad \dot{a}_0'' = \mathrm{Im}\left[1.5\dot{W}(A) + 0.5A\ddot{W}(A)\right]. \tag{18.55}$$

Example 3. The backslash nonlinearity, as plotted in Fig. 18.21a, has the harmonic linearization factor

$$W(A) = \frac{\alpha}{\pi} - \frac{\sin 2\alpha}{2\pi} - j\frac{\sin^2\alpha}{\pi},$$

$$\alpha = \frac{\pi}{2} + \arcsin\left(1 - \frac{2}{A}\right), \quad 0 \le \alpha \le \pi, \quad 1 \le A/t; \infty,$$

which means that

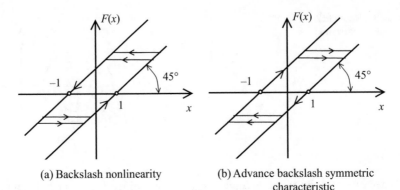

(a) Backslash nonlinearity

(b) Advance backslash symmetric
characteristic

Fig. 18.21 Backslash nonlinear characteristics

$$a_0 = \tfrac{1}{\pi}[(\alpha + \sin \alpha) - j(1 + \cos \alpha)]$$

$$a_0' = \tfrac{1}{\pi}(\alpha + \sin \alpha) \quad a_0'' = -\tfrac{1}{\pi}(1 + \cos \alpha)$$

$$r = \left|\tfrac{c}{2j}\right| = \left|j\tfrac{1+\cos\alpha}{\pi}e^{j\alpha}\right| = \tfrac{1+\cos\alpha}{\pi}.$$

The envelope equations come from Eq. (18.54):

$$u_0 = -(\alpha - \sin 2\alpha)/2\pi$$

$$v_0 = (1 - \cos 2\alpha)/2\pi$$

$$v_0 = 0.$$

Figure 18.22 plots the latter envelopes and the envelopes for the advance backslash symmetric nonlinear characteristic (see Fig. 18.21b) on the inverse Nyquist hodograph plane. The joint envelope of those characteristics

$$u_0 = -(2\alpha - \sin 2\alpha)/2\pi$$

$$v_0 = \pm(1 - \cos 2\alpha)/2\pi \tag{18.56}$$

separates the system robustness region (the unshaded zones in Fig. 18.22) over the class of nonlinear characteristics belonging the region

$$[F(x) + 1][F(x) - 1] \le 0 \quad \text{or} \quad F^2(x) \le 1. \tag{18.57}$$

This example illustrates that the first harmonic approximation assessment of robustness regions is feasible in the classes of systems where the conditions of Bongiorno [21] and Naumov–Tsipkin criteria are not met. The reduction of these sufficient conditions was obtained by M. Ostrovskii.

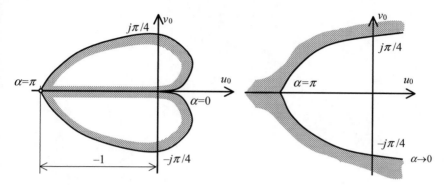

Fig. 18.22 Example 3. Envelopes of harmonic linearization curves on inverse Nyquist hodograph plane

At the same time, we believe that the biggest practical value of the harmonic approximation is the assessment of robustness regions in a dynamic system with particular nonlinearities in a given range of frequencies and amplitudes of actual signals.

References

1. Bishop D (1965) Vibration. Based on six lectures delivered at the Royal Institution, London in December 1962. University Press, Cambridge, 120 p
2. Chechurin SL (2014) Parametric resonance—pain and joy. St. Petersburg State Technical University, 68 p. Available at http://elib.spbstu.ru/dl/2/4853.pdf/view. Visited 20 Sept 2017
3. Butikov EI (2015) Simulations of oscillatory systems: with award-winning software, physics of oscillations. CRC Press, Boca Raton, p 363
4. Karlos R, Chechurin SL (1979) Analysis of the conditions for the excitation of parametric oscillations. Telecommun Radio Eng (English translation of Elektrosvyaz and Radiotekhnika) 33–34(8):93–95
5. Ostrovskii MY, Chechurin SL (1981) Parametric oscillations and stability of periodic motion in first harmonic approximation—1. Izvestia vyssih ucebnyh zavedenij. Priborostroenie 24 (10):20–29
6. Ostrovskii MY, Chechurin SL (1982) Parametric oscillations and stability of periodic motion in the first harmonic approximation—2. Izvestia vyssih ucebnyh zavedenij. Priborostroenie 25(7):27–32
7. Erikhov MM, Ostrovskii MY, Chechurin SL (1982) Synthesis of linear pulse systems using continuous model. Izvestia vyssih ucebnyh zavedenij. Priborostroenie 25(9):20–24
8. Murav'ev EI, Ostrovskii MY, Chechurin SL, Shmakov VE (1983) Parametric resonance in automatic control systems. Trudy LPI 391:42–45
9. Chechurin S (1983) Parametric resonance and stability of periodic motion. Leningrad State University, 219 p
10. Kapitza PL (1951) Dynamic stability of the pendulum with vibrating suspension point. Soviet physics—JETP 21, pp 588–597 (in Russian). In: Ter Haar D (ed) Collected papers of P.L. Kapitza. Pergamon, London, vol 2, pp 714–726, 1965
11. Chechurin LS, Merkoulov AI (2005) Robustness and describing function method. In: Proceedings of the IEEE International Conference Physics and Control, St. Petersburg, pp 393–398
12. Mandrik AV, Chechurin LS, Chechurin SL (2010) A method to stabilize the output signal of oscillatory system. Patent of Russian Federation, RF№2393520
13. Mandrik AV, Chechurin LS, Chechurin SL (2015) Frequency analysis of parametrically controlled oscillating systems. In: Proceedings of the 1st IFAC conference on modelling, identification and control of nonlinear systems MICNON 2015—Saint Petersburg, Russia, pp 24–26 June 2015, IFAC-PapersOnLine, vol 48, Issue 11, pp 651–655
14. Lee YK, Chechurin LS (2009) Conditions of parametric resonance in periodically time-variant systems with distributed parameters. ASME J Dyn Syst Meas Control 131 (3):68–79
15. Chechurin LS, Pervozvanski AA, Kim JH, Choi H (1998) Robust control in linear systems. SPbSPU, 203 p

16. Chechurin L (2008) Algebraic sufficient stability condition for periodical time-variant control systems (in Russian). In: Large scale systems control, №. 23. pp 24–38

17. Chechurin L (2008) Algebraic stability criteria for periodic time variant and nonlinear control systems (in Russian). St. Petersburg Polytechnic Univ J Eng Sci Technol Comput Sci Telecommun Control Syst 4(62):92–95

18. Chechurin L (2008) Harmonic stationarization method and evaluation of robustness of periodic time variant control systems (in Russian). In: Large scale systems control, №. 22, pp 70–85

19. Chechurin LS (2010) Frequency models and methods of analysis for dynamic system robustness. Dissertation for Doctor of Science degree, SPbSPU, 2010 (in Russian). Available at http://www.ii.spb.ru/admin/docs/Disser_Feb_01.pdf. Visited 20/09/2017

20. Chechurin SL, Chechurin LS (2003) Elements of physical oscillation and control theory. In: Physics and control, 2003. Proceedings of the IEEE international conference physics and control, St. Petersburg, 2005, vol 2, pp 589–594

21. Bongiorno JJ (1963) An extension of the Nyquist-Brakhausen stability criterion to linear lumped-parameter systems with time-varying elements. IEEE Trans AC 8:166–172

22. Chechurin SL, Chechurin LS (2005) Physical fundamentals of oscillation theory (in Russian), St. Petersburg State Polytechnic University, 258 p

23. Tchetchourine S, Hong SW (1997) Frequency analysis for dynamic systems. Computer Publishing Systems (CPS), St.Peteresburg

24. Leonov GA, Burkin IM, Shepeljavyi AI (1996) Frequency methods in oscillation theory. Springer Science+Business Media B.V., 403 p

Printed in the United States
By Bookmasters